Automate Testing for Power Apps

A practical guide to applying low-code automation testing
tools and techniques

César Calvo

Carlos de Huerta

BIRMINGHAM—MUMBAI

Automate Testing for Power Apps

Copyright © 2023 Packt Publishing

Group Product Manager: Alok Dhuri

Publishing Product Manager: Uzma Sheerin

Book Project Manager: Deeksha Thakkar

Senior Editor: Rounak Kulkarni

Technical Editor: Jubit Pincy

Copy Editor: Safis Editing

Proofreader: Safis Editing

Indexer: Rekha Nair

Production Designer: Rekha Nair

DevRel Marketing Coordinators: Deepak Kumar and Mayank Singh

Business Development Executive: Thilakh Rajavel

First published: October 2023

Production reference: 1131023

Published by Packt Publishing Ltd.

Grosvenor House

11 St Paul's Square

Birmingham

B3 1RB, UK

ISBN 978-1-80323-655-1

www.packtpub.com

To my wife, Gloria, and my children, Héctor and María, for supporting me and being a source of inspiration. To my co-conspirator in this endeavor, Carlos, whose companionship throughout this journey has really made the process more enjoyable.

– César Calvo

Low-code devs take flight,

Family, my guiding light,

Book blooms in moon's sight.

– Carlos de Huerta

Contributors

About the authors

César Calvo is a professional with more than 20 years of experience across the Microsoft technology stack and a blend of expertise across Microsoft Power Platform, Dynamics 365, Azure, and Microsoft 365. He has been a Microsoft Certified Trainer since 2008, with over 20 active official certifications in Power Platform, Dynamics, Azure, and M365. Recognized as a FastTrack Recognized Solution Architect – Power Apps in 2022 and 2023, he has found Microsoft Power Platform to be the perfect catalyst to create customer-centric solutions, specializing in the fusion code approach. César holds a degree in psychology and a computer science BSc degree.

Carlos de Huerta is a professional with over 20 years of experience in the ICT industry, where he has held various jobs, from business architect to advisor at Microsoft. He currently works there developing the technological strategy of the ecosystem of global partners and advisories. A telecommunications systems engineer from the Alcalá de Henares University, he acts as an ambassador in the community of architects and is an IASA Spanish board member. He participates with universities in the dissemination and teaching of the impact of technology on companies and their culture. Involved in the adoption of roles and techniques in the training field, Carlos has participated in the definition of cloud profiles with the Ministry of Culture in Spain.

About the reviewer

William de Bakker started as a developer 10 years ago and has worked extensively with Dynamics 365 and Power Platform technologies. He combines his critical thinking and business acumen to build elegant and efficient solutions for global deployments. His projects include building solutions for call centers that improve service, digitizing delivery and sales processes, and designing and developing applications.

Table of Contents

3

Power Fx Overview and Usage in Testing   55

4

Planning Testing in a Power Apps Project   75

Part 2: Tools for Power Apps Automated Testing

5

Introducing Test Studio – Canvas Apps Testing with a Sample Case Study

6

Overview of Test Engine, Evolution, and Comparison

7

Working with Test Engine   155

8

Testing Power Apps with Power Automate Desktop   181

Part 3: Extending Power Apps Automated Testing

9

PCF and Canvas Component Testing   211

10

ALM with Test Studio and Test Engine   233

11

Mocks with Test Engine   265

12

Telemetry and Power Apps   295

Index   319

Other Books You May Enjoy   328

Preface

Creating digital things to solve challenges is rewarding, but often, we lack the skills and find that the learning curve feels steep. Low-code platforms are changing this by making software development more accessible.

These tools are evolving to close the gap between a person's ideas and the code that brings them to life. However, there are essential techniques to understand, such as the *life cycle* of an app and the required *testing* to validate the expectations of your users. This book aims to equip you with these skills, especially in the context of **Microsoft Power Apps**, preparing you for the development of your next impactful app.

This book guides you through the process of testing in Power Apps, from the basic concepts to the best practice adoption in the creation of your apps. It will show you step by step, with clear examples, how low-code testing can help you build better applications and free you from annoying problems as you grow the features for your users. You will learn how to include automated testing in Power Apps canvas apps as well as in advanced areas such as PCF components and model-driven applications.

Who this book is for

Whether you are a citizen developer, professional Power Apps maker, or IT generalist who is interested in learning about testing automation to improve the business value and quality of your Power Apps, this book will help you to accomplish it. Working knowledge of Power Apps with a basic understanding of the Microsoft Power Platform is required to get the most out of this book.

What this book covers

Chapter 1, *Software Quality and Types of Testing*, introduces you to **Application Lifecycle Management** (**ALM**) and testing in the low-code landscape, emphasizing their role in enhancing quality, agility, and trust in business solutions.

Chapter 2, *Power Apps Studio Techniques and Automation Tools*, outlines the best practices for successful app development in **Power Apps**, focusing on testing techniques and a decision tooling matrix.

Chapter 3, *Power Fx Overview and Usage in Testing*, introduces **Power Fx** as the low-code language integral to the Microsoft Power Platform and highlights its role in test authoring and automation, aiming to enhance low-code productivity with current AI features.

Chapter 4, Planning Testing in a Power Apps Project, synthesizes prior knowledge for effective Power Apps development planning, emphasizing the role of testing in managing app evolution. It explores organizational **maturity models** to facilitate team-based development.

Chapter 5, Introducing Test Studio – Canvas Apps Testing with a Sample Case Study, introduces **Power Apps Test Studio** as a low-code testing solution, teaching you how to write and automate tests for canvas apps through expressions or recorders, illustrated with a practical gallery app example.

Chapter 6, Overview of Test Engine, Evolution, and Comparison, introduces the latest advancements in Power Apps testing, compares them with **Test Studio**, and guides you on combining both for optimal results.

Chapter 7, Working with Test Engine, guides you on using Test Engine, importing solutions into Power Apps, and aligning test plans with canvas apps.

Chapter 8, Testing Canvas Apps with Power Automate Desktop, introduces the key concepts of **Power Automate Desktop** and its application in UI testing.

Chapter 9, PCF and Canvas Components Testing, introduces the advantages of using Test Engine over Test Studio to automate the testing of canvas and **Power Apps Component Framework (PCF)** components, complete with practical examples.

Chapter 10, ALM with Test Studio and Test Engine, explores the fundamentals of **Azure DevOps pipelines** in relation to testing automation, detailing how to run canvas app tests using both the **Azure Pipelines classic editor** and **YAML**, as well as integrating tests built in Test Engine.

Chapter 11, Mocks with Test Engine, introduces the concept of using mocks in Test Engine for Power Apps, allowing test authors to simulate network calls and dependencies without altering the app or triggering external services.

Chapter 12, Telemetry and Power Apps, explores the importance of telemetry data in testing and debugging Power Apps, teaching you how to gather insights for more effective bug-fixing and user-centric testing.

To get the most out of this book

It is assumed that you have a working knowledge of Power Apps and that you are interested in increasing the business value and quality of your applications. Certain advanced areas of the book, such as PCF component testing would benefit from a professional Power Apps maker's knowledge; however, full details will be presented for you.

Software/hardware covered in the book	Operating system requirements
Test Engine	Windows, macOS, or Linux
Power Apps Test Studio	Modern browsers
Power Platform	Modern browsers
SharePoint Online	Modern browsers
Git	Windows, macOS, or Linux
Visual Code	Windows, macOS, or Linux

If you are using the digital version of this book, we advise you to type the code yourself or access the code from the book's GitHub repository (a link is available in the next section). Doing so will help you avoid any potential errors related to the copying and pasting of code.

Download the example code files

You can download the example code files for this book from GitHub at `https://github.com/PacktPublishing/Automate-Testing-for-Power-Apps`. If there's an update to the code, it will be updated in the GitHub repository.

We also have other code bundles from our rich catalog of books and videos available at `https://github.com/PacktPublishing/`. Check them out!

Conventions used

There are a number of text conventions used throughout this book.

`Code in text`: Indicates code words in text, database table names, folder names, filenames, file extensions, pathnames, dummy URLs, user input, and Twitter handles. Here is an example: "Gain insights into using essential test functions from the Power Fx reference, such as `SetProperty`, `Assert`, `Trace`, and `Select`, through examples."

A block of code is set as follows:

```
IfError( Collect( Students, { Name: txtInputName.Text } ),
    Notify("Invalid data provided. Please try again") )
```

When we wish to draw your attention to a particular part of a code block, the relevant lines or items are set in bold:

```
App.Formulas =
Distance = Velocity * Time;
Velocity = Value( txtVelocity.Text );
Time = sldHours.Value;
```

Any command-line input or output is written as follows:

```
setx MSBuildSDKsPath "C:\Program Files\dotnet\sdk\6.0.x\sdks"
```

Bold: Indicates a new term, an important word, or words that you see on screen. For instance, words in menus or dialog boxes appear in **bold**. Here is an example: "As an example, we have a simple calculator app with two labels for number input, one label for calculated results, and four **Add**, **Subtract**, **Multiply**, and **Divide** buttons."

> **Tips or important notes**
> Appear like this.

Get in touch

Feedback from our readers is always welcome.

General feedback: If you have questions about any aspect of this book, email us at customercare@packtpub.com and mention the book title in the subject of your message.

Errata: Although we have taken every care to ensure the accuracy of our content, mistakes do happen. If you have found a mistake in this book, we would be grateful if you would report this to us. Please visit www.packtpub.com/support/errata and fill in the form.

Piracy: If you come across any illegal copies of our works in any form on the internet, we would be grateful if you would provide us with the location address or website name. Please contact us at copyright@packt.com with a link to the material.

If you are interested in becoming an author: If there is a topic that you have expertise in and you are interested in either writing or contributing to a book, please visit authors.packtpub.com.

Share your thoughts

Once you've read *Automate Testing for Power Apps*, we'd love to hear your thoughts! Scan the QR code below to go straight to the Amazon review page for this book and share your feedback.

https://packt.link/r/1803236558

Your review is important to us and the tech community and will help us make sure we're delivering excellent quality content.

Download a free PDF copy of this book

Thanks for purchasing this book!

Do you like to read on the go but are unable to carry your print books everywhere?

Is your eBook purchase not compatible with the device of your choice?

Don't worry, now with every Packt book you get a DRM-free PDF version of that book at no cost.

Read anywhere, any place, on any device. Search, copy, and paste code from your favorite technical books directly into your application.

The perks don't stop there, you can get exclusive access to discounts, newsletters, and great free content in your inbox daily

Follow these simple steps to get the benefits:

1. Scan the QR code or visit the link below

https://packt.link/free-ebook/9781803236551

2. Submit your proof of purchase

3. That's it! We'll send your free PDF and other benefits to your email directly

Part 1:
Tools for Power Apps
Automated Testing

In this part, we will establish the concepts, tools, and processes needed for your testing journey. We will begin by discussing the significance of testing and application life cycle management. Then, we will go over the approaches and tools that will help us in practice, introducing Power Fx, the low-code language used across Microsoft Power Platform. This section concludes with a chapter on test planning in Power Apps projects, connecting the best practices discussed.

This part has the following chapters:

- *Chapter 1, Software Quality and Types of Testing*
- *Chapter 2, Power Apps Studio Techniques and Automation Tools*
- *Chapter 3, Power Fx Overview and Usage in Testing*
- *Chapter 4, Planning Testing in a Power Apps Project*

Software Quality and Types of Testing

1

Software quality is a critical aspect of any software development process, whether it's for traditional software development or low-code development under the citizen developer role. Testing plays a crucial role in ensuring that software solutions meet the quality and agility standards required for modern businesses. In this chapter, we will explore the concepts of **application lifecycle management** (**ALM**) and the **software development life cycle** (**SDLC**) and their importance in low-code development. We will delve into testing foundations, activities, and roles and examine how they help structure the testing process and contribute to maintaining healthy business processes, reducing time to market, and building trust in applications. Additionally, we will explore the various types of testing and the tester mindset necessary to achieve software quality in any app, from enterprise to small apps. This chapter aims to equip you with the knowledge and skills necessary to ensure software quality and speed while maintaining agility in today's fast-paced business environment.

In this chapter, we're going to cover the following main topics:

- Understanding how testing is part of the SDLC in low-code apps

- Exploring how ALM fits in testing low-code apps

- Examining the different types of testing and the mindset required for effective testing

- Discovering methodologies for the best Power Apps adoption, testing, and governance

By the end of this chapter, you will have gained an understanding of the critical role that software quality plays in modern businesses, and how testing is an essential part of ensuring this quality. You will have explored how ALM works for low-code apps, understand how testing is part of the low-code SDLC process, and know how maturity affects the level of adoption of those techniques.

Technical requirements

To test and follow some of the capabilities of this chapter, you will need a few technical requirements.

The material in this chapter will require a stable internet connection, and a compatible browser, such as Google Chrome or Microsoft Edge, will also be required to access Power Apps and other online resources, such as GitHub, which will be required to develop and test Power Apps. We will use a calculator app named Power Calc. Guidance and samples are located at `https://github.com/PacktPublishing/Automate-Testing-for-Power-Apps/tree/main/chapter-01`.

> **Important note**
>
> We recommend that you create a Power Apps Developer Plan so that you can test all functionality moving forward. You can follow the instructions at `https://powerapps.microsoft.com/en-us/developerplan/`, where you will get a free development environment to develop and test apps.

The need for testing for awesomeness and quality

Low-code platforms are a game-changer in developing a **minimum viable product** (**MVP**) as they significantly reduce the time and effort required to create an app. This efficiency does not necessarily mean that testing requirements are diminished, but it does enable the rapid transformation of identified business or personal needs into a functional application. This allows you or your targeted users to start using the MVP and experience its benefits in a relatively short period.

When apps start to grow or development goes too quickly, quality drops. To uphold this quality, testing is important. Although low-code platforms help democratize the development of applications so that everyone can transform their ideas into reality through software, there is still a gap where technology does not currently supply a seamless process to create and iterate those ideas. Users still need to understand some important concepts for successful application development until the vision outline in the preface allows everything to be automated just from a functional description of your needs.

Testing as a quality driver

Testing can help identify and fix defects and issues early on in the development process, which can help prevent delays and rework. Testing can help share expectations about how the app should work. Testing helps build better quality solutions and also allows you to keep. This can help improve collaboration and communication within the development team, which can lead to better-quality solutions.

So, better collaboration, better maintenance, and a reduction of time to market can lead to much more agility and speed to deliver an app.

For example, let's say you are creating an expense report low-code solution to manage employee expenses. Through testing, you may discover the following quality paybacks:

- Testing will help you validate requirements that have been gathered and defined in the ideate step. As you add an additional type of expense, the solution may not be able to integrate with the financial system. By identifying this issue through testing, you can review integration with the system to ensure that the expenses and financial records are consistent and up to date.

- The solution may not be able to handle different types of expense items. By identifying this issue through testing, you can implement additional logic and validation in the solution to ensure that the items are added and updated correctly.

- The internal team responsible for the expense delegation API wants to review how their service is used. You can reduce their time to review your app by sharing test use cases and test simulations of their functionality.

- As you add more features, you will check whether the previous functionality is still working to ensure you don't change data types, navigation, or anything that will negatively impact the user experience.

You do not have to be concerned too much about the possible implications of changes in each of the underlying components that keep the app functioning when you use a service such as Power Apps. However, you need to take care of changes to your app and the satisfaction of your users, and testing automation will help you save time and resources, enabling you to evolve your app more efficiently.

Software development life cycle

In this section, we will cover the **SDLC** standard for software development, review how it applies to low-code, and look at reasons why testing needs to be a central part of your process.

Figure 1.1 presents some typical stages in the SDLC process. It is a structured approach that enables you to create high-quality software at a reasonable cost. By following a step-by-step process, software development teams can design, develop, test, and effectively deploy software:

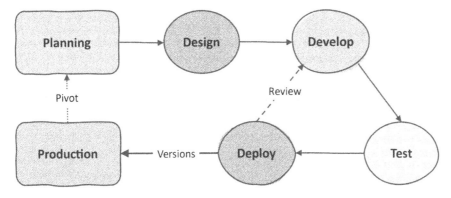

Figure 1.1 – The SDLC stages

Whether planned and managed or completely ad hoc, every step is expected. If you're producing software that people or just you use, you are following the SDLC inadvertently, and the main goal would be to embrace it formally (usually through a specific model such as Agile, Lean, or Waterfall, to name a few) and benefit from its adoption and related automation tools.

Let's briefly describe the activities involved when you develop a potential Expense Report Canvas App in Power Apps:

1. **Planning**: This step involves gathering requirements and analyzing them. First, requirements are gathered through user feedback, where a group of potential users is asked about the features they would like in the app – for example, the type of expense, receipt image recognition, or personal reports. Then, the collected requirements are analyzed to understand the problem better, figure out a solution, and make informed decisions. This analysis includes considering the gathered information, comparing existing similar processes for managing expenses, and determining what needs to be done to develop a more effective app.

2. **Design**: You start creating a plan for how the app will be made. In the SDLC, design is the stage where we figure out what our expense report app should do, how it should work, and what it should look like. We must think about all the different parts and how they will fit together to make a finished product that is useful and easy to use.

3. **Develop**: Based on the previous information and initial decisions taken, you begin the process of creating the software program. This involves defining the functionalities of the program, creating a design blueprint, writing the code, testing its functionality, and finally, releasing it for others to use. The develop phase is a crucial part of this process where you write the instructions for the computer to follow using Power Apps Studio.

4. **Testing**: From the definition of the app, this makes sure that your app works correctly. During testing, we try different things to see if the Expense Report Canvas App works as it should, and if it doesn't, we try to fix the problem so that it will work correctly.

5. **Deployment**: Once you have a working version of the app to use, you make it available to use. By deploying to the production environment – that is, publishing the Power App – the functionality will become available to end users.

As shown in *Figure 1.1*, based on the review of each version you develop, the cycle is repeated over and over again for every new version and functionality, going through the develop, test, deploy (and review) stages.

On the other hand, you could deploy the app in different environments based on the role of users using the app: production for final users to start creating and submitting their expense reports, staging for validating functionality with a specific group, or development while building a version. We will share more details about environments in this chapter.

As mentioned in the *Planning a Power Apps project* section at `https://learn.microsoft.com/en-us/power-apps/guidance/planning/app-development-approaches`, when we look at this process from a Power Apps perspective, it is accelerated thanks to the platform, and you can quickly create a new version of your app.

Figure 1.2 highlights this simplification in low-code with terms used in Power Apps and connecting them with SDLC stages. Here, **Design** includes the stages from requirements gathering to analysis and design, **Make** considers the development stage, and **Test and Run** reflects testing and running the app through fast iterations before deployment. Once you want to share with other users, **Publish** will make that deployment available in your environment of choice. We will map the terms with tools and capabilities later on:

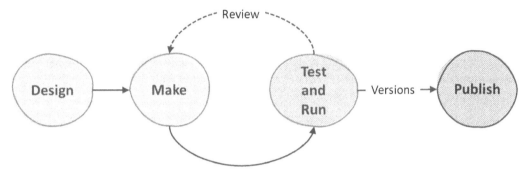

Figure 1.2 – The SDLC simplified for low-code

Although testing shows up as a stage in the SDLC, you should consider it not as a single stage but a whole process that expands across all stages of SDLC, from planning to development and production. This will ensure the **quality** of the app, as we described, but this will also help to bring an excellent experience to your users, giving the **awesomeness** wanted for our app.

Testing as an awesomeness driver

By testing the app, you can identify and fix any issues before the app is used by end users. When the need to fix a bug or defect arises, testing will help you identify the root cause, perform regression testing, validate the version in an environment, and then go live into production with confidence. When testing is included and automated in your development, fewer errors will occur.

You may wish to consider **inclusive design** in your app testing and development. More information can be found at `https://inclusive.microsoft.design/`, but in a nutshell, inclusive design guides you to create products that are psychologically, physically, and emotionally suitable for every person in the world, seeing human diversity as a resource for better designs.

Testing helps ensure that the app is user-friendly and provides a positive experience for end users, and by incorporating testing into the development process, organizations can build trust in the low-code solutions that are developed. This can help increase the adoption and usage of the solutions within the organization, as well as integration with existing development processes, by identifying and fixing defects and issues before the solution is released.

You can find some best practices for app design at `https://learn.microsoft.com/en-us/power-platform/developer/appsource/appendix-app-design-best-practices-checklist`. By providing readable names of controls, screen readers can read them out for blind people. You can create a Power Apps theme for consistency, and color accessibility or font uniformity.

So, a better code process, improved user experience, and increased trust lead us to a better-quality app and the process to deliver it.

For example, take the previous example of the expense report low-code solution to manage employee expenses. Empathy is an important part of design, so if we anticipate a disruption or improvement and advise our users about this, they will experience a better connection with the app. Through testing, you may discover the following improvements:

- The app may not follow the accessibility guidelines, and the app could be hard for people using screen readers. By reviewing the solution checker, and applying its recommendations, the developer can ensure that the solution will allow additional users to use it effectively.

- An update to the internal integration with the financial system is deployed with a change that affects the app. By identifying this issue through testing, the developer can alert internal users and the right stakeholders as changes are rolled back or a new version is published.

- The solution may not be user-friendly. By identifying users who take more time than expected to use a part of the app, through testing, the developer can redesign the layout and navigation of the solution to make it more intuitive and user-friendly.

- An error in the published app, due to a previous change, prevents users from successfully submitting a specific expense item. You confirm the test use cases didn't include this situation. It is updated and validated with a fix.

With that, we have described activities that are beneficial to improve the experience and quality of the expense report app. The way we described these activities implies a manual process. We can get the full benefit through automation processes, which is possible through the adoption of ALM. *Chapter 10* will look at ALM and test tools in more detail, but we will introduce these aspects in the next section.

Application Lifecycle Management (ALM)in low-code apps

ALM is a process that helps organizations manage the development, testing, deployment, and maintenance of software applications. In the context of low-code development, ALM can help ensure that the development of your low-code solutions is aligned with the overall goals and objectives of the organization, that they are developed and tested efficiently and effectively, and that they are released and maintained in a timely and controlled manner. ALM typically involves the activities highlighted in *Figure 1.4*:

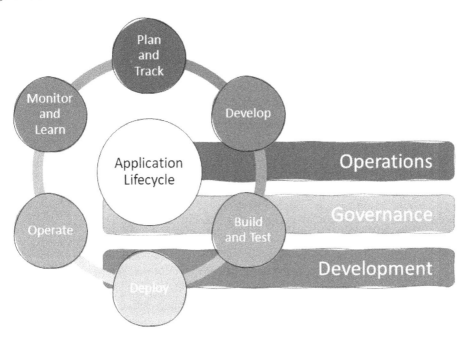

Figure 1.3 – ALM areas

Let's take a look at some of these areas in more detail:

1. **Application development**: This is where the low-code application is built and configured. It involves creating workflows, forms, and other components using the low-code platform's interface and customizing the application to meet specific business requirements.

2. **Maintenance and operations**: This involves ongoing support and maintenance of the low-code application after it has been deployed to end users. It includes tasks such as monitoring the application's performance, troubleshooting issues, and implementing platform release waves or weekly service updates.

3. **Governance**: This includes setting up guidelines and policies for the development, deployment, and maintenance of low-code applications. It establishes roles and responsibilities, security, and compliance requirements, as well as monitoring and auditing processes.

By expanding on the goals of each of these three areas, we can learn which Power Platform and Power Apps capabilities will help achieve them. The following list identifies the various Power Platform capabilities and tools for automation. It expands on the components, tools, and processes list available at https://learn.microsoft.com/en-us/power-platform/alm/basics-alm:

1. **Development lifecycle**:

   - **Environments**: In the context of Power Apps development, an environment refers to a specific instance of the Power App that can be used for development, acceptance, or production. This will allow us to define test environments versus production environments or for different life cycle scenarios. There are different types, such as Sandbox, Production, Developer, and Default.

   - **Solutions**: These refer to the packages that contain the components and configurations of a Power App. This will simplify advanced deployment and management. They will be the units to be deployed in environments.

   - **Source control**: This is a system that allows you to version control code and configuration files. It safely keeps and monitors changes to software assets, which is crucial when multiple developers work on the same files. It enables undoing changes or recovering removed files and supports healthy ALM by acting as the single access and modification point for solutions. You can use **GitHub** as a web-based platform as it allows developers to collaborate on software projects and automate continuous testing and deployment activities. It can be used to manage the code base of a an app. You can also connect with **Git**, the technology behind GitHub, from canvas apps: https://learn.microsoft.com/en-us/power-apps/maker/canvas-apps/git-version-control.

   - **Settings**: These are the parameters and configurations that control the behavior of the Power App. They allow you to activate features for debugging, monitoring, capabilities, or integration.

   - **Continuous integration and deployment**: Automating our testing and deployment processes for our app versions is critical to bring better quality and experience to our users. **Pipelines** refer to the process of automating the deployment of the Power App. This process can include tasks such as building, testing, and deploying the app to different environments. You can handle this process through DevOps services such as **Azure DevOps** or **GitHub Actions**, as outlined in *Chapter 10*.

2. **Maintenance and operations**: These refer to the tasks and procedures that are used to keep the Power App running smoothly, such as monitoring performance and fixing bugs. Logs refer to the records of the activity of the Power App, which can be used to troubleshoot issues:

   - **Monitor** is a tool that allows you to monitor the performance and usage of a Power App in real time.

   - **Application Insights** is a service that allows you to monitor and analyze the performance and usage of an application, such as an app developed with Power Apps. This provides a unique view of the application to the operations team through **Azure Monitor**, for example.

   - **Solution Checker** is a tool that allows you to check the solution's components, settings, and configurations so that you can identify and troubleshoot any issues. More information can be found at `https://learn.microsoft.com/en-us/power-apps/maker/data-platform/diagnose-solutions` and in *Chapter 2*.

3. **Governance**: The Power Platform Admin Center gives you tools to manage environments and security. It allows you to analyze the usage and performance of the platform, and it brings tools to manage environment roles or data and rights through **data loss prevention** (**DLP**) policies to avoid data breaches and protect data. The **Power Platform Center of Excellence (CoE) toolkit**, which will be reviewed in the last section of this chapter, includes guidance and tools for the best adoption.

> **Note**
> The adoption and implementation of all these components, tools, and processes will depend on the maturity of the organization to put the necessary processes, tools, and knowledge in place. In the following sections, you will learn more about how this can be done.

With that, we have reviewed ALM and its related Power Platform capabilities and tools. Now, it is time to deep dive into the concepts and practices for adopting testing successfully. First, we will review the activities or how you should identify what to test and who should be responsible for testing in the context of Power Apps and Power Platform.

As the platform evolves, it will include new automation capabilities, so we should adopt the new testing tools from the platform to simplify the app development. *Figure 1.3* shows the testing automation and tools that are available in the Power Apps ecosystem; we will look at these in more detail in the next chapter. In light colors, you can see low-code tools such as **Power Apps Test Studio**, **Solution Checker**, and **Monitor Tool**, and in dark colors, you can see tools for advanced scenarios where you can combine pro-code scenarios:

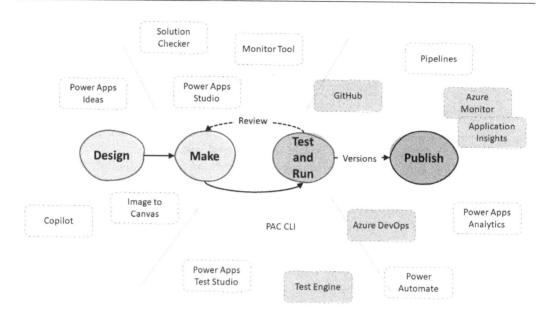

Figure 1.4 – Testing tools to automate and simplify testing and developer tools in the SDLC process

As a wrap-up, testing should focus on the external use of the app, its public components, and API dependencies, not on its internal execution. This will improve the customer experience and usage of your app.

Testing is your memory assistant. It's the best way to check the published app regarding its expected behavior. Testing acts as a reminder of how your app should work. The bigger and/or the older the app gets, the more complex it will be to find an issue or validate how it should work. So, adopting the automation testing tools and the process surrounding them will benefit you and your users.

Testing foundations

As we've discussed, testing refers to the process of assessing a software program to verify its behavior matches the program requirements defined. Many types of testing activities may be performed as testing foundations to validate this. These are the principles, concepts, and best practices that form the basis of software testing. These include things such as the importance of having clear and well-defined requirements, the need for thorough planning and coordination, and the importance of using appropriate tools and techniques to ensure the effectiveness of testing efforts. Let's start with the different activities and roles.

Activities and roles

Testing activities are the specific tasks and processes that are carried out as part of a software testing effort. These may include planning and controlling testing objectives and goals, analyzing and designing test cases, understanding the tools and services needed, and test implementation and execution. These activities help ensure that your low-code solutions are of high quality and meet the needs of the business. These activities may be carried out by different roles within an organization, such as software developers, testers, and **quality assurance (QA)** engineers. **Testing roles** are the different positions or jobs within an organization that are responsible for carrying out various testing activities, such as test leads and test engineers. As a citizen developer, you may also be involved in testing activities throughout the life cycle of your app. Each of these roles may have different responsibilities and expertise and may work together to ensure the success of the testing effort.

Once you are aware of the importance of adding testing to your app, one of the challenges is what to test and how to add tests. If you already have an application without any tests, how do you begin? Each app is unique, but you should consider the following guidelines to start with.

Prioritized and end-to-end flows, or the critical path

Focus on the top activities of your application. You want to check and validate that it is working as expected while you evolve and add new features for users, and on the other hand, that you find any issues users could face with the app. As part of the design stage, select the different parts that are the most valuable to the users. In our expense report app, think about the whole flow a user must go through to complete a job. Start creating test cases, following the flow from start to finish — from the main page, where you check the complete list of expense reports in our fictitious app — to the detailed information and the creation of one report — until you edit it and submit the expense.

Validate new features, one part at a time

As part of the new features you add to the app, review the main goal for the user, and break it down into steps so that you can, on the one hand, validate the expected result for the user, and, on the other hand, check whether future changes you may make will break the experience of the user.

From a best practice perspective, you should think of scenarios (you will map this to a test suite later) where specific features are validated as test cases to keep test cases small but group those tests with the same purpose.

In our expense report app, you may consider a test suite *new expense report*, where the user will follow several features and steps (test cases) to fulfill their objective – from report creation, expense creation, and editing to image receipts and expense report submission. This must also be done to validate that old functionality works as expected. An example of this includes submitting a special type of report for public sector companies.

Make it simple and fix it later, or the Boy Scout rule

The best part of the testing process is anticipating issues before your users face them. But once you or your users find an unexpected behavior, the best next step is to fix it and add tests around the bug to validate that it is working nicely.

When a new bug appears, it helps you understand why you didn't prioritize its flow or didn't add a test case specifically. In the design process, sometimes, you expect your users to use your app one way, but they end up using it in another way. Monitoring and feedback may give you important data from your app, such as a mismatch in expectations during development.

Finally, to fix a bug that one of your tests has detected, keep it simple and describe the expected outcome. The simpler and more assertive the test is about its expected result, the faster you will identify the coding issue. You will learn more about this when we cover the concepts of test expressions and test assertions in the next chapter.

Testing responsibilities

We may be the only one responsible for the app at the departmental level, or it may be part of an organizational-level app, and you could be part of a team involved in developing the app and expanding low-code and pro-code. Either way, testing is everyone's responsibility. However, the larger the scope of complexity of the app, the larger the team and roles' responsibilities. In these two scenarios, you could find the following:

- **Dedicated roles** in the scope of the whole organization or specifically for a project, including employees and/or external companies. In those teams, the maker/developer will be responsible for implementing the app and validating its behavior. You will find two levels in a software testing team:

  - **Test lead**: This person will be responsible for test planning, test governance, and coordinating with test engineers or testers

  - **Test engineers**: This person will be responsible for understanding what needs to be tested, developing and executing test cases, and test reporting

- Having a **single person** in personal development, departmental applications, or small businesses is usually the norm. The maker/developer is responsible for writing the code and ensuring that it works as expected. Having said that, all people involved should participate in the testing process through some of the mentioned activities.

Now that we have reviewed the activities and roles that help us identify what to test and who should be responsible for testing, let's drill down into the mindset of the tester.

The mindset of a tester

Among the various factors that contribute to successful testing, the psychological aspect holds a significant position as it can influence the way we approach testing without our conscious awareness. This can be attributed to several reasons:

- A solution-oriented mindset versus a problem-oriented approach tends to be less effective when it comes to testing code

- It can be challenging to identify defects in something that has been created by yourself

- It can be difficult to consider potential issues when the focus is on what the system should do

Testers do not typically need to have a deep understanding of how the system under test works. Instead, they need to adopt the perspective of the end user and consider potential scenarios from the user's point of view. In this context, your knowledge of how the system works can prevent you from identifying alternative scenarios that may lead to unexpected behavior.

Therefore, to be an effective tester, you need to focus on identifying ways to break software. A software tester's job entails not only finding bugs but also preventing them. This includes analyzing the requirements, process optimization, and implementing a continuous testing strategy. In this sense, a tester's mindset entails being concerned with quality at all stages of the SDLC. Because quality is the responsibility of the entire team in agile development, the primary focus of agile testing is shifted toward the initiative and controlling activities that prevent the occurrence of defects.

This connects us to the following three key areas of agile development, in which it's acknowledged that testing is not an isolated stage but an essential component of software development, together with coding, thereby summarizing a consolidated view of many of the capabilities for testing:

- **Mindset**: This is where everyone is responsible for ensuring quality and running tests as a cross-process and not just a phase through customer collaboration and thinking in terms of requirements elicitation

- **Skill set**: In the role of tester, as a low-code developer, think of adopting skills to do different types of testing through automation and effective communication and collaboration

- **Toolset**: Use development and build tools for the best performance and use examples and requirements as guidance and support (visual examples, mockups), as well as simplification (recording, multi-level test automation in fusion teams, and so on)

Agile testers must depart from the guiding concepts and operational procedures of conventional software development. Success as an agile tester requires the appropriate mentality. Twelve principles can be used to summarize the agile testing mindset, as shown in *Figure 1.5*:

Figure 1.5 – Agile testing principles

Let's take a closer look at them:

1. **Quality assistance over quality assurance**: Quality assurance is the process of ensuring that the software meets certain quality standards before it is released to the customer. Quality assistance, on the other hand, is the process of helping the customer achieve their quality goals by providing guidance and support throughout the development process.

2. **Continuous testing over testing at the end**: Continuous testing is the process of testing software throughout the development process, while testing at the end is the process of testing the software only once it is completed. Continuous testing allows for the early detection of defects and allows for faster delivery of the software to the customer.

3. **Team responsibility for quality over the tester's responsibility**: In traditional testing approaches, the responsibility for quality lies solely with the tester. However, in a whole-team approach, the responsibility for quality is shared among the entire team, including developers, testers, and other stakeholders.

4. **Whole team approach over testing departments and independent testing**: A whole team approach is a process that involves all members of the team in the testing process, including developers, testers, and other stakeholders. This approach allows for better communication and collaboration among the team members and leads to a more efficient and effective testing process.

5. **Automated checking over manual regression testing**: Automated checking is the process of using automation tools to test software, while manual regression testing is the process of testing software manually. Automated checking is faster and less prone to errors than manual regression testing.

6. **Technical and API testing over just UI testing**: Technical and API testing is the process of testing the underlying technical aspects of the software, such as the code and the APIs. UI testing is the process of testing the **user interface (UI)** of the software.

7. **Exploratory testing over scripted testing**: Exploratory testing is the process of testing software by exploring it, without a preconceived test plan. Scripted testing is the process of testing software by following a predefined test plan. Exploratory testing allows for a more flexible and creative approach to testing and can lead to the discovery of defects that may not have been found through scripted testing.

8. **User stories and customer needs over requirement specifications**: User stories and customer needs are the processes in testing software that consider the needs and wants of the customer, while requirement specifications test software by following a predefined set of requirements. User stories and customer needs allow for a more customer-focused approach to testing and can lead to a better test to satisfy the purpose of the feature developed.

9. **Building the best software over breaking the software**: Building the best software is the process of creating software that meets the needs of the customer and is of the highest quality, while breaking the software is the process of finding defects in the software. Building the best software allows for a more customer-focused approach to testing and can lead to a better understanding of the customer's needs.

10. **Early involvement over late involvement**: Early involvement is the process of involving all members of the team, including testers, early in the development process, while late involvement is the process of involving testers only at the end of the development process. Early involvement allows for better communication and collaboration among the team members and leads to a more efficient and effective testing process.

11. **Short feedback loop over delayed feedback**: Having a short feedback loop is the process of providing feedback to the team members promptly, while delayed feedback is the process of providing feedback to the team members only after a significant amount of time has passed.

12. **Preventing defects over finding defects**: Preventing defects is the process of identifying and addressing potential issues before they occur, while finding defects is the process of identifying issues after they have occurred. Preventing defects allows for a proactive approach to testing and can help minimize the number of defects that are found in the software.

Finally, it is important to understand the different types of testing and how they apply to development.

Types of testing

Some common types of testing in development include performance testing, unit testing, integration testing, system testing, acceptance testing, and UI testing. Using the principles of **inclusive design**, we should consider accessibility or localization testing as well. Unit testing focuses on testing individual components or units of your low-code solution, while integration testing focuses on testing how the components work together. System testing focuses on testing the end-to-end functionality of your solution, and acceptance testing focuses on verifying that your solution meets the needs of the business. By understanding these different types of testing, you can develop a test-driven mindset and ensure that the low-code solutions you develop are of high quality. Let's discuss each type in more detail.

Unit testing

The goal of unit testing is to identify and fix any issues with individual units of the application before they affect the overall functionality of the system. This can help ensure that the application works correctly and meets the specified requirements. You should consider it when you develop **Power Apps code components**.

Integration testing

This is a type of software testing that is used to evaluate the interfaces between the different components of an application or system. This type of testing is performed to ensure that the different components of the application are working together properly and meeting the specified requirements. You perform this when you use third-party connectors or when you build a *custom connector* and want to validate the integration.

System testing

The goal of system testing is to identify any issues or defects that may affect the overall functioning of the app from a feature perspective (functional) and a performance, security, or scalability perspective (non-functional). This may involve creating a specific **Power Platform environment** with your solution test data and the needed integrations or connectors in a test environment.

Acceptance testing

This type of testing is typically performed by the end user or a representative of the end user and focuses on evaluating the overall functionality and performance of the application from the user's perspective. We will explore this in detail in *Part 3, Planning a Power Apps Project*.

Exploratory testing

Exploratory testing is an approach to software testing that emphasizes creativity, learning, and adaptability. It involves testing a software product without a formal test plan or script and relies on the tester's intuition, experience, and skills to discover issues and opportunities for improvement. It may involve using your app while running **Power Apps Monitor** watching out for errors, app performance issues, accessibility or design problems, and unexpected error messages.

UI testing

This is a type of software testing that focuses on the visual aspects of an application, such as its layout, design, and UI. UI testing may be performed manually, by having a tester visually inspect the application and compare it to the visual specifications, or it may be automated, using specialized tools such as **Test Engine** or **Power Apps Test Studio** to compare the actual behavior of the application to the expected behavior.

Overall, each of these testing processes is important in its own way, and they all play a crucial role in ensuring the quality and success of a software application. By performing these tests at different stages of the development process, it is possible to identify and resolve any issues with the application before it is released, which can help improve the user experience and ensure the success of the app and the happiness of your users.

Now, it is time to start the final section of this chapter and review how all the previous content comes together.

How should I apply the theory?

The **Customer Advisory Team** (**CAT**) within Microsoft Power Platform engineering comprises solution architects whose primary objective is to aid customers in expediting the adoption of Power Platform. The next section will present two of the main outcomes of their latest work. By introducing the Maturity Model and the Center of Excellence accelerator, you will see important elements for the organization for adoption and testing activities.

Power Platform Maturity Model

Through working closely with some of the platform's most accomplished users, the CAT team has discerned recurring patterns, practices, behaviors, and themes that enhance the progression of thriving organizations as they embark on a comprehensive digital transformation journey with Power Platform. This exercise was based on **Capability Maturity Model Integration** (**CMMI**). It is a framework for process improvement that provides organizations with a set of best practices and guidelines for managing and improving their processes. It is used to evaluate an organization's current processes and to identify areas for improvement, and it covers a wide range of process areas, including project management, engineering, and service management. You can check team content on their YouTube channel: `https://aka.ms/powercatlive`.

The result of this work is the Power CAT Adoption Maturity Model, which defines five stages based on the maturity level of the organization. This won't be a static view, but a journey where each organization will adopt new capabilities and processes, moving up to the adoption of the platform:

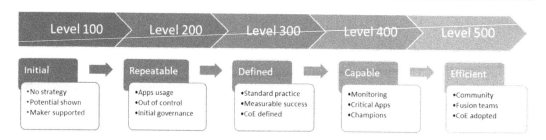

Figure 1.6 – CAT Adoption Maturity Model levels summary

As we can see, the first two levels (100 and 200) are both early stages in the adoption of Power Platform within an organization. Both stages may have a lack of consistent strategy and governance, with the use of Power Platform seen as *out of control* until administrative and governance controls are put in place.

Level 300 and beyond refer to the advanced stages of Power Platform adoption within an organization. At this stage, the focus is on standardization and achieving measurable success with Power Platform. The organization follows standardized processes for managing and monitoring Power Platform that are automated and well understood by makers. At Level 500, an organization has successfully demonstrated Power Platform's ability to transform critical capabilities quickly and effectively.

These levels offer significant benefits for testing as they provide a well-defined, standardized, and automated process for testing, which ensures the quality of the developed apps and flows and also make it easy to validate the impact of the platform on the organization.

The CMMI presents a broader view from different points of view: strategy and vision, business value, admin and governance, supporting and nurturing citizen makers, automation, and fusion teams. In the next section, we will review the Center of Excellence toolkit, which is based on CMMI, along with tools that will be part of your testing strategy adoption.

We'll look at this framework again in *Chapter 4* as we will use it to plan a testing phase in Power Apps.

The Power Platform Center of Excellence toolkit

The Power Platform CoE toolkit is a set of tools and resources that organizations can use to establish a CoE for Power Platform, as part of the CMMI methodology. The goal of the CoE is to provide a centralized approach that covers all aspects, including governance, adoption, community engagement, development, operations, security, and data management. We have summarized and listed some for you to review:

- **Governance, adoption, and community**:

  - Guidance and templates for establishing governance policies and procedures for Power Platform, such as user access and permissions, data management, and compliance

  - Resources for promoting the adoption of Power Platform within an organization, such as training materials, user adoption plans, and best practices

- Resources for building and nurturing a community of Power Platform users, such as user groups, forums, and events

- **Development life cycle, solution management**:

  - Guidance and templates for managing the development life cycle of Power Platform solutions, such as source control, continuous integration, and testing

  - Tools for managing and maintaining solutions built on Power Platform, such as a solution checker, solution templates, and solution packages

- **Operations, maintenance, and security**:

  - Tools and best practices for monitoring and troubleshooting Power Platform solutions, such as log analysis, performance monitoring, and incident management

  - Guidance and best practices for securing Power Platform solutions, such as data encryption, access controls, and compliance

- **Data management**: Guidance and best practices for managing and protecting data used by Power Platform solutions, such as data backup, data retention, and data archiving

The toolkit is a great source of guidance overall, specifically for testing adoption. *Figure 1.7* highlights four tools that can accelerate some of the activities mentioned previously:

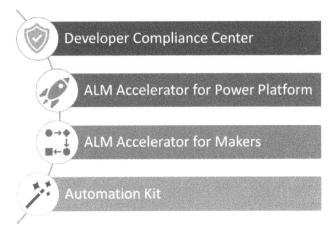

Figure 1.7 – Tools from the Power Platform CoE toolkit

From top to bottom, you can verify compliance with your app, automate publishing solutions between environments, manage version control and deployment, or implement an automation platform while following industry best practices. Check out the toolkit to learn more.

The Expense Report app

We have talked about a fictitious app throughout this chapter. You can download and build the app for yourself at `https://learn.microsoft.com/en-us/power-apps/maker/canvas-apps/expense-report-install`. How will you use the different components, tools, and processes with this app? Based on the CI/CD example from `https://learn.microsoft.com/en-us/azure/architecture/solution-ideas/articles/azure-devops-continuous-integration-for-power-platform`, we will look at the app, as shown in *Figure 1.8*, so that you can use it as a personal project while reviewing the different elements. In this example, looking into the CMMI model, the organization would be at Level 300: Defined, where there is an environment strategy and ALM is facilitated:

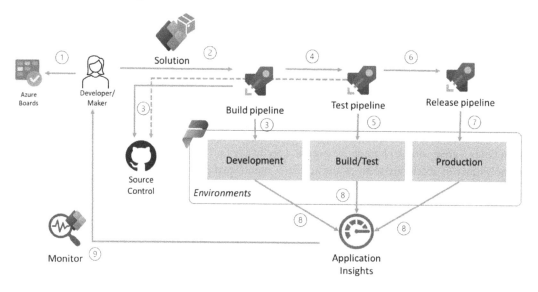

Figure 1.8 – CI/CD architecture for Microsoft Power Platform

Let's look at the steps involved in more detail:

1. In the planning phase, requirements from users are created in Azure DevOps Boards as a way to track the functionality developed.

2. The solution for the app is updated as part of the CI/CD process. This triggers the build pipeline.

3. Continuous integration exports the solution from the development environment and commits files to source control. This allows us to track the source code that's deployed in each environment. Test cases will also be stored in the source control repository.

4. Continuous integration builds a managed solution, runs tests, and creates a build artifact.

5. You deploy to your build/test environment. Tests created with Test Studio are executed using Test Engine or the PAC CLI.

6. Continuous deployment runs tests and orchestrates the deployment of the managed solution to the target environments.

7. You deploy to the production environment, making the app available to the final users.

8. Application Insights collects and analyzes health, performance, and usage data.

9. You review the health, performance, and usage information. You could use a monitoring tool to check performance and app behavior as well.

This brings us to the end of this chapter.

Summary

This chapter covered the crucial role of software quality in modern businesses, emphasizing the importance of testing for both traditional and low-code development. It presented insights into the SDLC and ALM and their significance in low-code environments. This chapter delved into various testing types, the tester mindset, and methodologies for adopting and governing Power Apps. At this point, you should have a comprehensive understanding of how to ensure software quality and agility, maintain healthy business processes, reduce time to market, and build trust in applications so that you're equipped with the skills to thrive in today's fast-paced business landscape.

In the next chapter, we will review Power Apps' built-in capabilities and automation tools to help you debug, troubleshoot, and test your apps.

Power Apps Studio Techniques and Automation Tools

In the previous chapter, we looked at the importance of testing and application life cycle management as critical fundamentals of software development. Now, we need to consider the techniques and tools that support us from a practical perspective. We will describe best practices and recommendations for successfully developing an application in Power Apps in the context of testing.

This chapter will help you understand the tooling and capabilities available in Power Apps, and how they can help you to better develop and test apps. This is achieved through steps such as debugging an issue, monitoring a performance problem once the app has been deployed, commenting and collaborating for the best teamwork, or analyzing your apps to understand how the users run them. Many of them will be reviewed deeply through this book, but here, you will get comprehensive insights into the tooling available and its built-in capabilities. A visual representation of the tools and features will be shared for the best understanding and decision-making.

In this chapter, we're going to cover the following main topics:

- Understanding debugging and learning the best practices to debug Power Apps using in-app tools such as Tracing or Monitor
- Getting familiar with Power Apps capabilities to accelerate app development using App Checker, and to help understand and review your apps.
- Learning features to test your apps through built-in techniques and tools
- Getting an understanding of the testing automation tools available for Power Apps, such as Test Studio, Test Engine, and EasyRepro, and which apply when

By the end of this chapter, you will have developed a better understanding of the various Power Apps techniques and automation tools available to assist you with app development and testing. You will have learned how to debug Power Apps using built-in capabilities from Power Apps Studio such as Checker or Monitor Debugging, and gained familiarity with the process to help test apps by understanding what the code does. *Figure 2.1* presents a stack view of them so that you can see how they relate to each other:

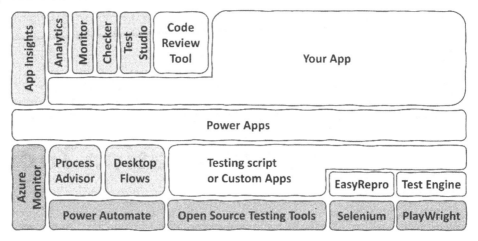

Figure 2.1 – Stack view of Power Apps tools and features

With these skills, you will be well equipped to improve your Power Apps development and testing processes and achieve greater efficiency in your app development projects.

Technical requirements

To test and follow some of the examples inthis chapter, there are a few technical requirements.

For this chapter, you will require a stable internet connection. A compatible browser, such as Google Chrome or Microsoft Edge, will also be required to access Power Apps and other online resources, such as GitHub, which are required to develop and test Power Apps. We will use a calculator app named **Power Calc**. Guidance and samples are located at `https://github.com/PacktPublishing/Automate-Testing-for-Power-Apps`.

> **Important note**
>
> We recommend that you use Power Apps Developer Plan so that you can test all functionality moving forward. You can follow the instructions at `https://powerapps.microsoft.com/en-us/developerplan/`, where you will get a free development environment to develop and test apps.

Debugging in Power Apps testing – from why to how

As we discussed in the previous chapter, the purpose of software testing is to identify any defects, bugs, or problems in your apps before they are released to end users so that they can be fixed or addressed promptly. Testing helps improve the software's quality and increases the system's reliability, usability, and performance. However, once you get an alert or error from your testing cases, you will need to fix it – that is, you will need to start debugging your app.

Debugging is the process of finding and fixing errors or bugs in your apps. It is an important part of software development as errors or bugs can cause a program to produce unexpected results or stop working altogether. It helps ensure that an app is working as intended and meets the requirements and expectations of your users once a problem has appeared, the same way testing helps you in the first place.

You may go through some simple steps:

1. Study your app code and available data from your tests, or data from a user sharing the problem.

2. Review what behavior was expected and why you got the unexpected result.

3. Understand whether there are more related cases. Did you consider the type of data the user wanted to add? Did you expect the user to use it that way? What else should be considered?

4. As part of the fixing process, you will include new tests, or change test cases to validate it, and then after successfully running your app, make it available to the users.

These are your basic debugging steps. Effective debugging requires three main skills: a good understanding of the code, knowledge about the capabilities of the development environment, and the ability to think logically and systematically. Debugging can be a time-consuming and challenging process, but it is essential for delivering high-quality software that meets the needs of users.

To debug your Power Apps code, you can use the built-in debugging and tracing capabilities of Power Apps. These features allow you to view the flow of your app, as well as any errors or warnings that may occur. You can also use the Monitor tool to track your app's performance and identify any potential issues.

We will go through the three skills we mentioned previously – that is, getting a good understanding of your code, gaining knowledge of the built-in tools in the development environment, and using data to systematically resolve those bugs and optimize your code:

Figure 2.2 – Debugging pillars

Next, we will dive into better understanding your code.

Understanding your code

Check out the documentation as a starting place for Power Apps troubleshooting: `https://learn.microsoft.com/en-us/troubleshoot/power-platform/power-apps/welcome-power-apps`. We will only focus on the key elements to help you understand your code, as well as techniques to help you quickly resolve any issue using Power Apps Studio. It is important to interpret what Power Apps is telling us about our code and the elements we have in the app, as well as how we can adopt best practices for better troubleshooting.

Let's review some of them in detail.

Results view

In Power Apps, the results view allows you to see the results of a specific query, formula, or property object and where you can check the expected data type format. This view can be used to validate expected data from a data source such as a SharePoint list, an Excel workbook, or a Dataverse entity. *Figure 2.3* highlights two important elements: the formula bar, for presenting the values, operators, and functions used for the corresponding property, and the results view. In this case, we can see the welcome message text label in the Power Calc app:

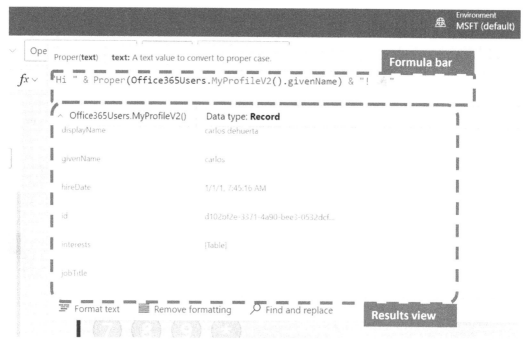

Figure 2.3 – The formula bar and results view for the welcome message text

We positioned the cursor in the `MyProfileV2()` method; the results view allows us to check that it returns a record data type so that we can see which data is available. This helps us better determine the `givenName` selection. On the other hand, when you are building the app and you get an error, you can check the formula bar and the provided message to validate what is going on. In *Figure 2.4*, the **Power Apps editor** is complaining as the `Proper` function requires a `Text` value and `Office365Users.MyProfileV2()` returns a record data type:

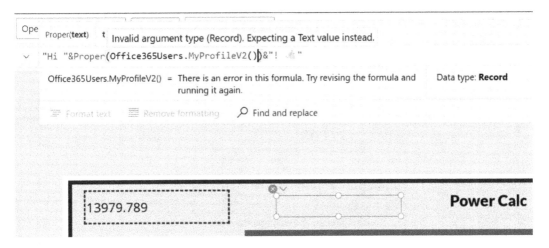

Figure 2.4 – Formula bar error

In general, you can navigate to a gallery, collection, or variable of interest to see its value from the results view for better debugging.

Finally, as part of the editor experience, IntelliSense shows you code for the potential properties or data names. This area will appear as you start typing or after an object or function with valid values (see *Figure 2.5*):

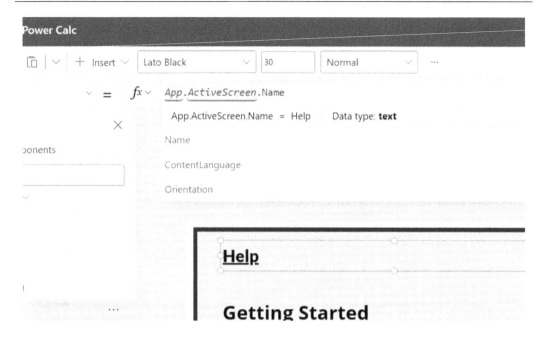

Figure 2.5 – IntelliSense shows properties for ActiveScreen

Naming conventions

Make it your routine to change the names of screens and elements as you create them. As your app becomes more complex, your loss of control will increase over which element belongs to a specific feature, which will result in you spending more time in the debugging process. Default-named elements can cause unnecessary extra work when you're trying to find the right element every time you fix a function or want to try something new. Hence, maintaining a uniform naming convention for all objects in your Power Apps application is crucial, such as the ones published in the Power Apps Canvas Coding Standards and Guidelines at https://powerapps.microsoft.com/en-us/blog/powerapps-canvas-app-coding-standards-and-guidelines/. Adherence to a defined guideline will help you develop and debug your apps with the benefits presented in the following table:

Principles	Benefit
Simplicity	By following a consistent and simple naming convention, developers can easily understand the purpose and functionality of different elements within the code. This reduces the cognitive load and makes the code more maintainable, as there is less ambiguity in interpreting the code's behavior.
Readability	Naming conventions that prioritize readability ensure that code can be quickly understood by anyone who reads it, even if they didn't write it. This promotes better collaboration among team members and allows for quicker understanding of code base structure and functionality, reducing the time needed to onboard new developers.
Supportability	A well-structured code base with clear naming conventions allows for easier debugging and support. If a problem arises, developers can more quickly identify and fix the issue, resulting in increased uptime and user satisfaction. Here, you will add documentation and collaboration details as well.
Testability	Clear naming conventions and adherence to best practices make the code more testable. Developers can more easily identify what each part of the code is supposed to do, making it simpler to write effective tests. This, in turn, leads to more reliable code and a better end-user experience.
Ease of deployment and administration	Naming conventions and best practices that consider deployment and administration can streamline these processes. For instance, having consistent naming for configuration files, scripts, and other deployment-related components simplifies automation and reduces the likelihood of human error. This leads to quicker and more reliable deployment and maintenance of the application.
Performance	Following best practices can lead to more efficient code execution, from simple recommendations of periodically publishing your apps to benefit from platform optimization, to using a local collection for small datasets or understanding delegation when managing data sources, resulting in a faster and more responsive application.
Accessibility	By adhering to best practices that include naming conventions for accessibility, developers ensure that the application can be used by as many people as possible, including those with disabilities. People using screen readers will better navigate and understand what the app is doing and how to use it. This expands the potential user base and may also fulfil legal requirements in certain jurisdictions.

Table 2.1 – Some benefits of best practices and naming convention adoption

We will come back to some of these principles later in this chapter, such as accessibility and performance. For now, let's close the first skill!

Control helper

This is a simple trick to check some functionality and data as you develop your app. It uses a control to help you test or visualize a specific action or data. Sometimes, you will want to validate the results of a function just because you think the collection or variable holds some data it does not, so you could then add a button and bind the function to try it or see the output of some function or record through the app's usage. Adding a label right in front of you to show a variable that you depend on helps you better understand the flow of your code. These activities shouldn't persist in your application code as they are usually part of the development flow.

If you just want to check global variables or collections, go to the **Variables** menu and navigate through the different options, as shown in *Figure 2.6*. Here, you can see where global variables are set and how, as well as collection data or additional variables:

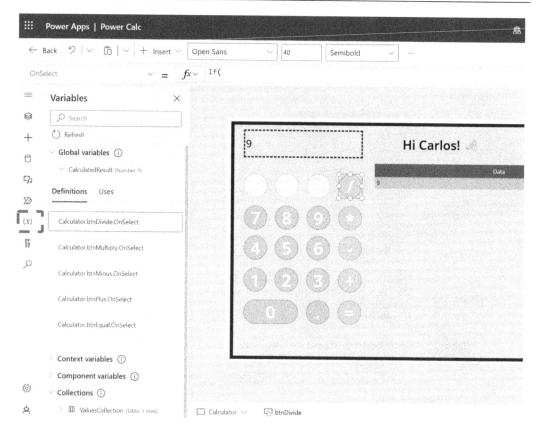

Figure 2.6 – The Variables menu option and the Definitions view

Having these features for understanding code, let's continue with the next step in debugging.

Development environment

One of the most powerful built-in features of Power Apps for debugging is the **Monitor** tool. With the help of Monitor, makers may diagnose and resolve issues by viewing a stream of events from a user's session. Canvas app developers can utilize Monitor to examine events while creating a new app in Power Apps Studio or keep an eye on published apps while they are running. Model-driven app developers can track key actions such as form-related errors, page navigation, command executions, and other critical actions to better understand app activity.

> **Note**
>
> Check out *Chapter 12* for a deep-dive review of Monitor and Application Insights, and the information available from the Monitor logs. To debug your app in production, it is recommended that you use Application Insights to gather any information needed for troubleshooting.

Monitor and Trace

Using Monitor, you can build more reliable apps and diagnose and resolve issues more quickly. It gives you a detailed understanding of your app by capturing all the significant activities that take place while it is in use. To enhance performance and spot any faults or issues, Monitor also gives you a better understanding of how the events and algorithms in your app function. We will focus on canvas apps, although you can check out the related information for model-driven apps in the documentation at https://learn.microsoft.com/en-us/power-apps/maker/model-driven-apps/monitor-form-checker.

If you want to track a specific value as part of your code, you should use the `Trace` function. It will record information behind the scenes to track specific data or comments you want to associate with any action or function. For example, to track the value of the `ValuesCollection` calculator app collection, you must add the following code to the button when an error occurs:

```
Trace( JSON ( ValuesCollection, JSONFormat.Compact), TraceSeverity.
Information);
```

First, from Power Apps Studio, click on the left menu named **Advanced tools**. Then, select Open monitor, as shown in *Figure 2.7*, and check the trace results:

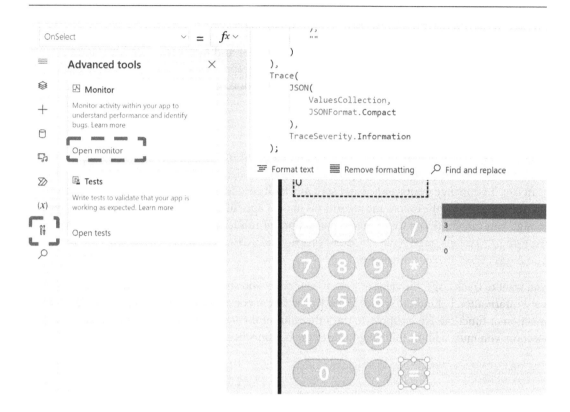

Figure 2.7 – Trace logging when clicking a button

Once you open the **Monitor** dashboard and connect to the current session of your canvas app, as soon as you or the user start browsing and using your app, events will flow into the Monitor tool.

Using the app, we tried to divide by zero. Due to this, an error occurred, and the data regarding `ValuesCollection` is available in the **Monitor** panel, as shown in *Figure 2.8*:

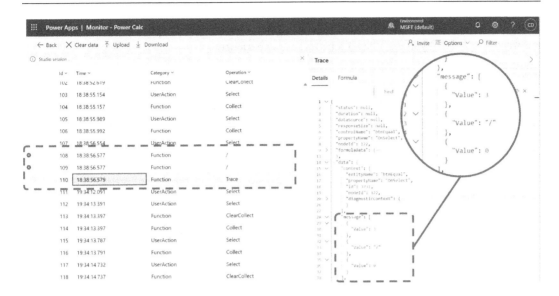

Figure 2.8 – Trace log in the Monitor panel

The fundamental principle of debugging an issue is to gain a deeper comprehension of what your application accomplishes and how it does so. Merely examining the application formulas or runtime errors is sometimes insufficient to isolate a problem. Observing the sequence of events as they transpire within your application can help elucidate the progression of events and how your application is functioning, enabling you to detect errors and diagnose problems more rapidly.

Here are five examples of problems and queries that you can uncover when using Monitor:

1. **Performance issues**: Monitor can provide insights into which connectors, functions, or components are taking the longest time to execute, identify slow-running queries and network connection issues, and help you optimize your app's performance.

2. **Errors and exceptions**: Monitor identifies errors and exceptions that occur in your app. It can provide details on the error message, stack trace, and other relevant information, allowing you to quickly diagnose and resolve the issue.

3. **Data inconsistencies**: Monitor can help identify data inconsistencies in your app, such as data that is missing or not displaying correctly. It can provide insights into the source of the issue, allowing you to take corrective action.

4. **User behavior**: Monitor can provide insights into which features are most popular, which screens users are spending the most time on, and other valuable information that can help you improve your app's usability.

5. **Security issues**: Monitor identifies security issues in your app, such as unauthorized access. It can provide details on the source of the issue, allowing you to take corrective action to secure your app and protect your data.

By using the Monitor tool, you can better understand the behavior of your app, and after considering these areas as examples, you can deliver a better user experience.

Error handling

As you develop your app, you may anticipate potential problems through the use of expressions to detect errors and to provide an alternative value or action. For example, you can use `IfError` to test whether a certain function contains errors, and then display an error message with `Error`. You can also use the `IsError` function to check for errors in your formulas and handle them appropriately.

> **Important note**
>
> To use these error-handling functions and expressions, you will need to activate the formula-level error management preview feature by ensuring **Settings** | **Upcoming features** | **Preview** is turned on.

In addition to using built-in features and formulas, you can use custom validation rules to ensure that the data that's entered into your app is accurate and complete. There are several ways to validate data in Power Apps. One approach is to use the built-in validation features of Power Apps, such as the **Required** and **Numeric** properties for input fields. These properties can be used to ensure that certain fields are filled out and that the data entered is of the correct type. Some examples include the following:

- The `Validate` function determines whether a value is valid based on the information from the data source – for example, `MaxLength` of column data

- The `IsMatch`, `Match`, and `MatchAll` functions test whether a string matches a pattern

- The `IsNumeric` function examines whether a value is numeric

The following table presents some examples of its use in the expense app:

Formula	Description	Result
`Validate(Expenses, PersonalSplit, 120)`	Checks whether 120 is a valid value for the `PersonalSplit` column in the `Expenses` data source.	"Values must be between 0 and 100."
`IsNumeric(txtEmployeeID.text)`	Check if the employee ID is a number (for example, AB4114)	false
`IsMatch("carloshm@company.com", Email)`	Matches an email address	true

Table 2.2 – Validation functions

While debugging from a code perspective and in a development environment is essential, analyzing data is equally important for ensuring that your app is functioning as expected and meeting business requirements. Let's move on to the next step to gain insights.

Process and data analysis

In Power Apps, telemetry data refers to the data that is collected about how users interact with the app, such as how often they use certain features, which screens they visit most frequently, and which types of devices they use to access the app.

The data presented in the Monitor tool we referred to in the previous section is telemetry data, which we could use to analyze how our app performs as well. For example, in *Chapter 12*, you will learn how to import Monitor logs for query analysis; you will also learn how to create and configure Application Insights. You can use **Application Insights**, which is a cloud-based service that helps you monitor and analyze the usage and performance of your app, to analyze telemetry data.

An example of the potential capabilities of the Application Insights service alongside Power Apps can be seen in *Figure 2.9*. **Kudos App**, available at `https://learn.microsoft.com/en-us/power-platform/solution-templates/hr/employee-kudos/install`, presents the different navigation patterns the users of the app have followed from the dashboard screen to other screens:

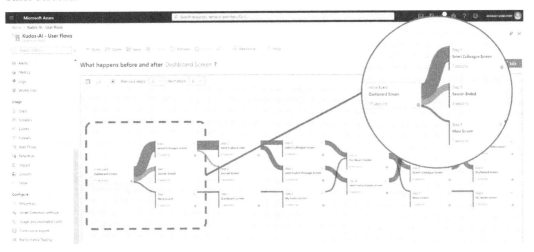

Figure 2.9 – App insights for Power Apps (this screenshot intends to
show the layout; text readability is not important)

Note that collecting telemetry data can help you improve your app's performance and user experience, but you must respect users' privacy and comply with any applicable laws and regulations. You should always provide clear and transparent information about what data you collect and how you use it, such as any personal information you share and explicitly trace and the information presented by the service, as you will see in *Chapter 12*.

Process Advisor

Imagine you could use your telemetry data to understand what is going on with your app so that you could visualize its actions and success. Imagine a tool that could optimize the flow process we saw previously for Kudos App. Process Advisor is a tool in the Power Platform that helps organizations analyze and improve their business processes. As a Power Apps developer, you can use Process Advisor to identify inefficient or outdated processes and optimize them to increase productivity and efficiency.

Process Advisor will analyze your process and then provide recommendations for process improvements, including suggested actions and insights into the potential impact of those actions.

To use Process Advisor, you need to create a process map that captures the steps and activities involved in a particular business process. You could create a process map by importing data or Power Automate recordings (check out *Chapter 8*), but we will benefit from a Power Apps insights template (`https://learn.microsoft.com/en-us/power-automate/process-mining-templates#power-apps-insights-template`) published to simplify this process. In this case, you will need the Application Insights ID associated with your Application Insights service.

Once you have created your process map, you can use Process Advisor to analyze it and provide recommendations for improvements. These recommendations may include automation opportunities, such as using Power Automate to automate repetitive tasks or optimizing data-handling processes to improve accuracy and efficiency.

As an example, you could use the **CaseID** field name with the **AppID** property from Application Insights, as well as the **Activity** field name and the message from Application Insights from the documentation:

- The **Case ID** field is used to uniquely identify each case or record that is created as part of a business process. The value that's stored in the **Case ID** field is unique and helps keep track of the history and status of each case.

- The **Activity** field is used to track the current stage or step of a case within a process. This field is often used to identify what actions need to be taken next for a case and to ensure that cases are being progressed through the correct steps.

The use of a process map allows you to visualize and gain valuable insights into the processes used in your app. By examining a graphical representation of how different app actions are performed, you can identify areas where opportunities for improvement exist.

It might take several minutes for the analytics results to appear, but an example of the KPIs and visualizations is presented in *Figure 2.10*:

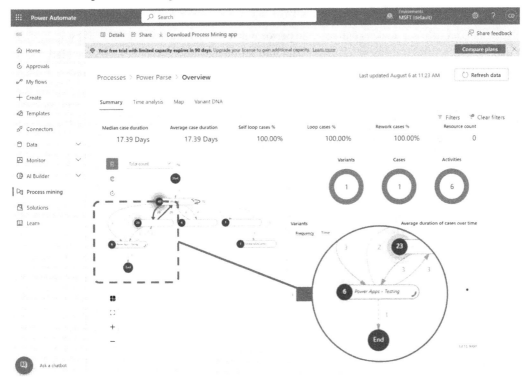

Figure 2.10 – Process Advisor analytics

The previous figure represents a process map, where you can check the time duration for each step and flow activity so that after multiple aggregation variants of users using your app, you can optimize it. For example, we added a Power Apps testing message trace for when we're running tests in Test Studio or Test Engine (the node at the bottom left) so that we can find activities the user has completed and we didn't include.

App and administrator analytics

Is there an out-of-the-box capability in Power Apps to analyze usage and performance, or even a country/region view of that usage? The answer, as you may expect, is that you have several options based on your role. We will view Power Apps analytics at the app level in *Chapter 4*, and at the tenant level in *Chapter 12*, but we want to highlight them as key out-of-the-box analytics tools.

In the Microsoft Power Platform admin center, environment admins can access analytics that offer insights into usage, errors, and service performance at the environment level. These reports can be leveraged for governance and change management services to improve the user experience.

Analytics features help app makers enhance the performance and comprehension of apps by analyzing user data. These features transform telemetry data into actionable tasks and offer recommendations to improve your apps. Canvas apps analytics and performance insights (`https://learn.microsoft.com/en-us/power-apps/maker/common/performance-insights-overview`) for model-driven apps are self-service tools for app makers that analyze runtime user data and provide a list of recommendations to help improve the performance and provide a better understanding of apps. To start using them, follow these steps:

1. Go to the Power Apps portal.
2. Open **Apps** and select any canvas a or model-driven app.
3. From the contextual menu or command bar, select **Analytics (Preview)** for canvas apps or **Performance (Preview)** for model-driven apps.

As we continue our journey through Power Apps techniques for better testing, we want to emphasize the importance of comments and documentation. While debugging code and analyzing data, it's essential to have a clear understanding of what's happening at each step. Comments and documentation can help you keep track of your thought process and make it easier for you to revisit your code later. It may seem like an extra step, but it will save you time and energy in the long run and improve your collaboration with others. Let's get started!

Collaborative notes, comments, and documentation in your apps

Although the intention of the formulas you added in your app, or the different elements used to build it, might be clear after some time, you may find this less clear in the future, and even forget why it was done that way. This information is crucial in fixing issues or delivering new features to your app.

Some features will help you have a better understanding of your code or that of others. We will discuss these features in the next few sections.

Debugging together

There is a nice functionality to collaboratively see a **Monitor** session or even connect to a published app so that makers and support teams can watch the sequence of events generated by the user's interaction with your app:

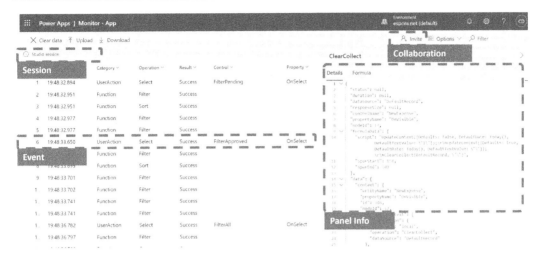

Figure 2.11 – Monitor areas for collaboration

Once you launch a **Monitor** session, you can click **Invite** and add a user to the same session. *Figure 2.11* shows some of the main elements:

- **Collaboration**: This menu option allows you to invite a user via their name or email

- **Event**: Anyone joining the session will see all events except the ones available before sharing

- **Session**: The main session (Studio session) will provide all the available options, but the invitation (guest session) will not be able to change the source filter and will access the same **panel info** data

Code comments

Code comments are a useful way to document your Power Apps code and make it easier to understand and support. In Power Apps, you can add comments to your code by using the double-slash (//) symbol for line comments, as shown in *Figure 2.12* and in the following code:

```
// This is a comment to clarify some code
```

You can create a block comment that spans multiple lines by using the special key combination of a forward slash and an asterisk (/* */). This block comment will continue until you explicitly terminate it with the opposite key combination of an asterisk forward slash:

```
/*
This is s multiline comment to clarify your code:
Example1: Content
*/
```

Check the following screenshot as an example of this:

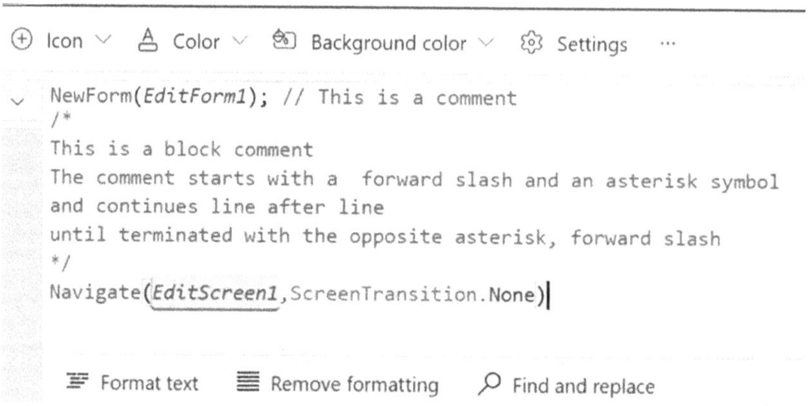

```
⊕ Icon ∨    Ⓐ Color ∨    🖼 Background color ∨    ⚙ Settings    ⋯

∨   NewForm(EditForm1); // This is a comment
    /*
    This is a block comment
    The comment starts with a  forward slash and an asterisk symbol
    and continues line after line
    until terminated with the opposite asterisk, forward slash
    */
    Navigate(EditScreen1,ScreenTransition.None)|


    ≣ Format text    ≣ Remove formatting    🔍 Find and replace
```

Figure 2.12 – Block comment

You can add comments to your Power Apps code at any point, but it is often helpful to include them before or after blocks of code that may be difficult to understand and review them after changing the code as well.

Collaborative notes and user mentions

With collaborative notes and user mentions, you and your team can review your app and add feedback to improve the experience or any implementation details you would like to be changed. For example, let's say that early in the development process for a feature of your calculator app, you wanted to add feedback about the math functions to be used. You can create a comment and mention-tag people in the comment by using @+person's name so that they can see the proposed app features.

> **Important note**
>
> This is a preview feature in canvas apps at the time of writing. Preview features are not intended for use in a production environment and may have limited functionality. These features are made available before an official release to give customers early access and allow them to provide feedback.

Here's what collaborative comments look like in canvas apps. Here, you can add comments at every level in the app elements:

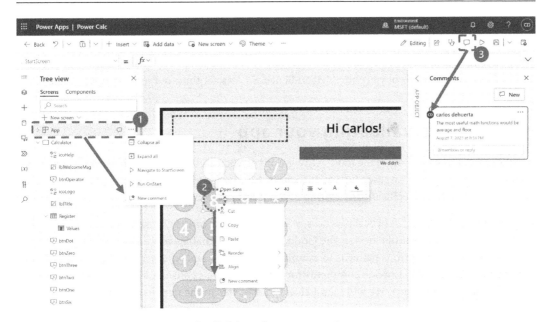

Figure 2.13 – Collaborative comments in canvas apps

In all examples, when you provide feedback or comments for you or others, follow these simple rules for the best clarity:

- Write comments to describe the goal of the code or the purpose you think is missing from your testing. Avoid repeating what can be understood from the code.

- Comments are part of the app. Change them as needed, and use commenting actions to edit, delete, or resolve feedback.

- Avoid future rework – that is, write self-descriptive code with key clear comments that don't require external references (for example, acronyms).

> **Note bonus**
>
> There is an experimental feature that uses Git version control to edit canvas apps. This allows you to use GitHub collaboration features and develop in a more pro-code environment such as Visual Studio Code. If you explore this scenario, you could install a free extension from the Visual Studio Marketplace called **CodeTour**. It allows you to record and play back guided walk-throughs of your code.

Now that we've emphasized the significance of feedback and comments in the development of Power Apps, it's time to explore ways to improve and validate your app. In the next section, we'll delve into different techniques and tools to ensure that your app meets the necessary quality standards and is well suited to meet the needs of your intended audience. So, let's continue our journey toward creating top-notch Power Apps!

Improving and validating your app

As mentioned in *Chapter 1*, one of the most active groups at Microsoft for building tools and publishing best practices is the Power **Customer Advisory Team** (**CAT**). They create many community-based open source tools. One that is specifically interesting in this context is the Power Apps Code Review Tool.

With its customizable checklist of best practices, a 360-degree view of app checker results, app settings, and a free search code/formula viewer, the Code Review Tool enables more efficient app reviews. The checklist consists of various patterns to examine in your app, and whether it meets the specified criteria or not. You can also provide comments for the app creator's consideration or explore further information about the pattern. We will take a closer look at this in the next few sections.

Before you publish and go live with your app, you should use available tools such as the **solution checker** or the patterns from the **Code Review Tool** to test for potential issues, better performance, accessibility, and more. Let's review them.

Solution checker and App checker

Using the solution checker, you can perform a comprehensive static analysis check on your solutions. This involves comparing them against a set of best practice rules to swiftly identify any problematic patterns. Once the check is complete, a comprehensive report is generated, outlining the detected issues, impacted components and code, and links to documentation that provides solutions for each issue. This applies to model-driven apps; however, the test is available through the **App checker** feature in canvas apps. The **App checker** feature lists any formula errors, accessibility warnings, and performance optimization tips, as shown in *Figure 2.14*. You can run it by clicking on the stethoscope icon in the top-right menu of Power Apps Studio. As you adjust your app, the errors or suggestions in the stethoscope are updated:

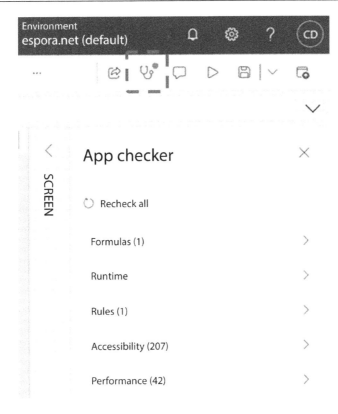

Figure 2.14 – App checker in canvas apps

If you have packaged your canvas app in a solution, the **Power Apps** checker can provide you with a list of potential issues from the solution checker in the **Power Apps maker portal**.

However, to view a summarized list of issues, it is necessary to republish any existing apps with the most recent version of Power Apps Studio. These steps are shown in *Figure 2.15*, where you can see the different issues presented in the app selected:

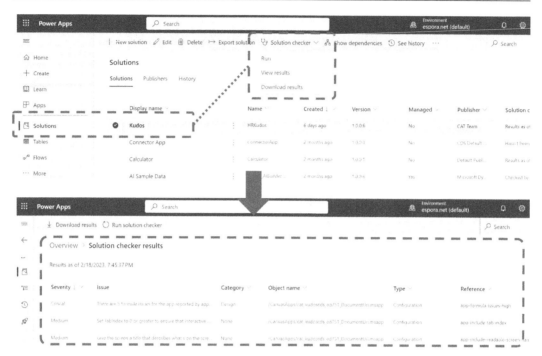

Figure 2.15 – Solution checker results

When you're looking for an automated way to integrate your DevOps processes, a more advanced form should be used, such as the **PowerShell module**, an **Azure DevOps task**, or direct web **API interactions** (https://learn.microsoft.com/en-us/power-platform/alm/checker-api/overview).

The **App checker** tool runs a set of rules to find errors and show potential issues when you're running your app. It covers the following fields:

- **Formulas**: Are there missing quotes or invalid references?
- **Runtime**: Will the app pose a reliability risk or is the app using deprecated functionality?
- **Rules**: Will conditional rules create issues, such as preventing delayed loading (this feature is deprecated)?
- **Accessibility**: Will the app create issues for keyboard or screen reader tools?
- **Performance**: Are there any potential performance problems in the app?

Now, let's move on to the next tool.

Power Apps Review Tool

Power CAT has done a great job. First, they created components that you can use for an excellent user experience with the **Power CAT Creator Kit**, available at `https://github.com/microsoft/powercat-creator-kit`. This is a toolkit that helps you create a well-designed experience on the web and mobile. The Power Apps Review Tool uses it; you can follow the steps in the GitHub repository to explore it.

The Power CAT Creator Kit is a comprehensive set of resources, tools, and best practices designed to enable Power Apps users to effectively plan, design, develop, and deploy high-quality applications. The toolkit offers a range of features, including customizable templates, guidance documents, deployment scripts, and sample code, which help users streamline their development workflows and achieve better outcomes.

Its customizable templates and guidance documents are tailored to different use cases, such as custom connectors, model-driven apps, and canvas apps. Additionally, the toolkit offers extensive documentation on best practices and common issues, as well as deployment scripts that automate the process of deploying apps to various environments. As a valuable resource for any Power Apps user, we will present two of the most important in the context of testing: **Review Checklist** and **Code Viewer**.

Review Checklist

Review Checklist provides a series of best patterns that you check in your application. For this example, we uploaded the Kudos app from the **Employee Experience Starter Kit**, available at `https://powerapps.microsoft.com/en-us/blog/powerapps-employee-experience-starter-kit/`. It presents each pattern, its description, and details from the app, as shown in *Figure 2.16*. If we drill down into the first pattern, **N+1 Database or API requests**, it will look into N+1 queries. For example, when within a gallery, `LookUp()` or `Filter()` operations on a data source are bound, triggering too many requests to servers:

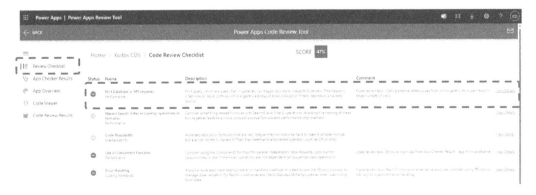

Figure 2.16 – Review Checklist

Clicking on the **View Details** option will show the result of the analysis, along with detailed information on the specific screen, control, and property involved in the issue and a link to the documentation:

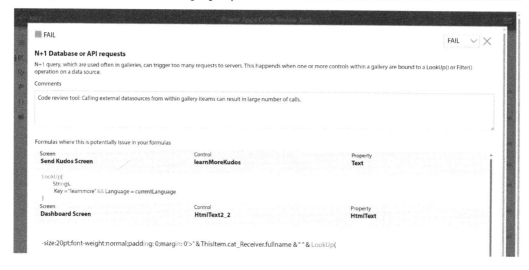

Figure 2.17 – N+1 Database code review result

Code Viewer

Code Viewer lets you read through the code associated with each screen and control to add comments and assign best practice patterns manually. *Figure 2.18* highlights the actions where, after selecting an element in the code browsing icon, you can add available best practices to be considered in that code from the documentation or personal experience:

Figure 2.18 – Code Viewer features

Finally, the information from **Code Viewer** and **Review Checklist** is presented through the **Code Review Results** report. It helps you understand where you should start making changes and how those changes will affect the score of your app. An example from the app we've imported is shown in *Figure 2.19*. A summary of the failed patterns is presented and categorized by severity and area so that you can prioritize its resolution and improve the experience of your users:

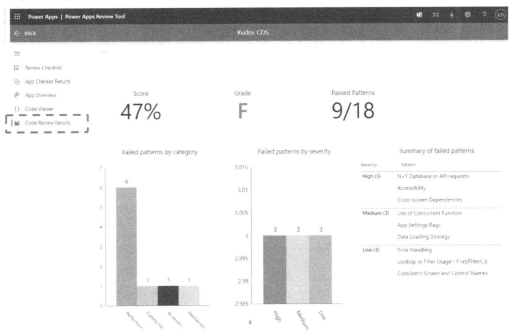

Figure 2.19 – Code Review Results

Overall, there are several ways to validate the code in your apps, and the approach you choose will depend on the specific requirements of your app. By using the built-in validation features, formulas and expressions, and best practice patterns, you can ensure that your app will work as expected and investigate if it doesn't while gaining insights into how your app is used.

Exploring the Power Apps testing automation tools matrix

So far in this chapter, we have shared available best practices and tools for better debugging your app and resolving issues that are found while you develop and check the app for your users. Those recommendations are used throughout the entire ALM process, as presented in *Chapter 1*, to streamline the testing process. These are presented in *Table 2.3*. The different tools are classified to guide you on when and why to use each based on your expertise and the state of the app:

> **Note**
>
> We've reviewed some of these tools in this chapter; we will discuss the rest of them in upcoming chapters and dive deeper into many of them, specifically Test Studio and Test Engine.

	Power Apps Expertise	Canvas app Support	Model-driven Support	Stage	Maturity
Power Apps Studio built-in	Low	Yes	Yes	Make, Test and Run	High
Test Studio	Low	Yes	No	Test and Run	Medium
Test Engine	High	Yes	Yes	Test and Run	Low*
Power Automate desktop flows	Medium	Yes	Yes	Test and Run	Medium
EasyRepro	High	No	Yes	Test and Run	None*
Solution Checker	Low	Yes	Yes	Make	Medium
Code Review Tool	Medium	Yes	Yes	Make	Low*
Monitor Tool	Low	Yes	Yes	Make	Medium
Application Insights	Medium	Yes	Yes	Publish	High
Analytics	Low	Yes	Yes	Publish	Medium
Process Advisor	High	Yes	Yes	Design	Low

Table 2.3 – Testing automation tools matrix

The matrix includes four different attributes for each tool:

- **Power Apps expertise level**: This helps you understand which ones you should start with so that you can master them and move to more complex ones that you can set up and include in your learning journey:

  - **App Makers** should focus on **Low** for tools that are part of the Power Apps Studio out-of-the-box experience, and **Medium** for those that require or use cross-platform tools, along with their installation and external dependencies

  - **Pro Developers** should focus on **High** for advanced tools as they let them learn and connect with pro-code tooling and development while expanding low-code capabilities

- **Canvas apps and model-driven app tool support**: Although this book focuses on canvas app development, the two models converge uniquely while preserving their unique features. For example, Test Engine includes model-driven apps due to EasyRepro being deprecated. This allows you to see how the platform has evolved for common behavior in this space.

- **Stage**: As we have reviewed, testing activities expand to all phases and not just the development process. This will help you understand when you should look into those tools for the best impact on your app. It is not a hard classification as you may use them in different phases:

  - **Ideate**: This tool will help you better understand your app's internals for a new version.

  - **Make**: The information provided by the tools and features will help you optimize and resolve issues before your users face them.

  - **Check & save**: As you prepare to publish or test initial versions of the app, these tools will help you validate expected behavior and alert you when changes break it.

  - **Publish**: This focuses on gathering live information from your users so that you can identify potential real scenarios that might not be considered. This information usually goes into the Ideate phase as input.

- **Maturity**: Last, but not least, as a personal non-official view on the relevance of each tool, we classify them based on the perceived evolution of the tools. So, you could include their adoption based on your current expertise, the maturity of the tool, and its importance in the ecosystem:

  - Those marked with an asterisk (*) indicate an open source project in GitHub. This means that no matter the adoption of the community, you can use it and extend it on your own if it suits your needs. However, the more adoption a project gets, the more updates will be needed to fix issues or extend features.

  - **None**: Tools that, although available, didn't have recent updates.

  - **Low**: These are features or tools that are well integrated into the platform and have not changed recently or that could be used with your apps but do not have a specific integration scenario (desktop flows).

  - **Medium**: Features with recent changes, even in preview, are to be considered in your prioritization.

  - **High**: These are features and tools with high activity and are fast-evolving, such as AI capabilities. This area could include the team responsible for the feature or tool's capabilities, such as Test Engine, and you could explore and experiment with it in advanced scenarios.

To close this chapter, let's look at *Figure 2.20* from the bottom (platform services or open source tools) to the top (Power Apps capabilities); this provides a stacked view of the tools presented earlier. If we look below Power Apps, we'll see dark-colored boxes, which indicate core platform services. The other boxes are features of those core platform services or open source apps that extend to Power Apps testing capabilities. When we look above Power Apps, the dark-colored boxes extend their capabilities, while the light-colored boxes are apps developed in the platform. This view helps us understand the role of each service in the platform, as well as their integration capabilities and use:

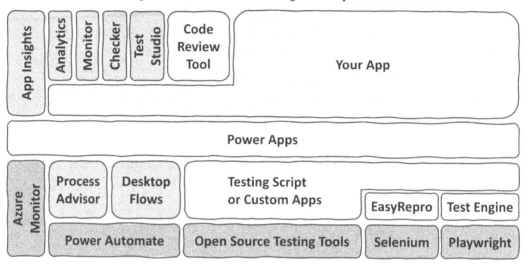

Figure 2.20 – Stack view of tools and features

The Power Platform and, specifically, Power Apps are rapidly evolving. You can check this out by looking at the available testing tools. **EasyRepro** started as a library to help customers run UI testing for model-driven apps. **Test Studio**, on the other hand, was added to canvas apps as an integrated tool to create and run functional tests. Finally, **Test Engine**, as a testing automation tool for model-driven apps and canvas apps, is bringing the capabilities of Test Studio to close the loop for an end-to-end DevOps adoption. In Test Engine, you can run tests recorded with Test Studio and automate them in tools for continuous integration and delivery, such as **Azure DevOps** or **GitHub**. These tools can be integrated with Power Apps to automate various aspects of the testing process, such as running tests, collecting results, and reporting on issues. *Chapter 10* will present them in the context of ALM. You could use other open source testing tools to build apps or automation targeting the Power Apps platform. Examples of third-party testing tools include testing frameworks and libraries such as **Power Automate desktop flows**, **Puppeteer**, **Playwright**, **Selenium**, **Cypress**, and **xUnit**, among others. The downside is that you will lose the added value of existing extensions that target Power Apps, such as the ones we will use throughout this book. *Chapter 5* to *Chapter 8* will go through Test Studio, Test Engine, and Power Automate in more detail.

With this, we have come to the end of this chapter.

Summary

In this chapter, we reviewed several best practices for developing applications in Power Apps from a testing perspective. We explored the importance of debugging and how to test Power Apps using in-app tools such as Checker, Tracing, or Monitor, while highlighting the significance of collaboration in the development process, and how to achieve this through proper commenting and documentation.

To turn data from our app into useful information, several tools and capabilities were presented. This opened up a whole new perspective on how to visualize the information with Application Insights' features, Process Advisor's upcoming integration, and open source applications shared by Power CAT.

Lastly, we looked at the Power Apps testing automation tools matrix as a way to contextualize the different features and testing automation tools available for Power Apps, such as Testing Studio, Testing Engine, and EasyRepro.

In the next chapter, we will talk about Power Fx – the low-code language that's used across the Power Platform – and how it is used for test automation, focusing on its use in Power Apps canvas apps.

3

Power Fx Overview and Usage in Testing

In the previous chapter, we delved into the essential aspects of debugging, monitoring, commenting, and tracing in Power Apps development. We explored built-in capabilities and various automation tools to optimize the app development process. The key topics we covered included best practices for debugging using in-app tools. Additionally, we discussed built-in techniques and tools for app testing and the different automation tools.

In this chapter, we will introduce you to Power Fx, the low-code language that's utilized across Microsoft Power Platform. You can use it broadly in Power Apps, but we will share specific examples and apply them to Power Apps testing. We will explore AI preview features available at the time of writing and guide you through the following main topics:

- **From spreadsheet to Power Fx**: Learn how to transition from Excel examples to using Power Fx in Power Apps while navigating the language reference

- **YAML – the backbone of Power Fx formatting**: Understand YAML's role in formatting Power Fx for Power Apps

- **Exploring the top test functions in Power Fx**: Gain insights into using essential test functions from the Power Fx reference, such as `SetProperty`, `Assert`, `Trace`, and `Select`, through examples

- **Natural language to Power Fx – harnessing the power of AI**: Discover new capabilities that enable you to develop using natural language so that you can add Power Fx code and programming by example

By the end of this chapter, you will be able to navigate the Power Fx language, understand the YAML formatting for Power Apps, and apply test functions such as `SetProperty`, `Assert`, `Trace`, and `Select` effectively. Furthermore, you will see how the platform is adopting natural language capabilities. These skills will empower you to improve your Power Apps development and testing processes, leading to increased productivity in your app development projects.

Technical requirements

To replicate the examples in this book, you must download the corresponding files from the designated GitHub repositories. We will utilize the GitHub repository specifically created for this book at `https://github.com/PacktPublishing/Automate-Testing-for-Power-Apps/tree/main/chapter-03`.

Access to a functioning Power Platform environment is necessary to use the provided examples or create new ones. The Power Platform developer environment appears to be the most suitable solution for this purpose, as presented earlier in this book. On the other hand, some examples will run without any dependencies. You will find an application in the repository called `ConsoleREPL.exe` that you can download and run with the examples that, due to the open source nature of Power Fx, will run with its own engine.

From spreadsheet to Power Fx

Imagine crafting an app with the same ease as creating an Excel worksheet. What if you could harness your existing spreadsheet expertise for app development? Power Apps and Power Fx were born from these questions, aiming to empower millions of Excel users to build apps, automation, virtual agents, and more using familiar concepts.

Power Fx is an expression-based language, similar to other programming languages that use expressions to represent calculations. We will describe an example with the basics of the language but consider that it is a general-purpose functional programming language. You can create variables (`velocity` or `time` would be your variables), use operators (in this example, `*` operator multiplication), and any functions (such as the `Round` example to round up if the next digit is 5 or higher). For instance, the following expression represents the multiplication of `velocity` and `time`:

```
Velocity * Time
```

However, Power Fx takes this concept a step further, turning ordinary expressions into meaningful formulas.

In Power Fx, you create formulas that bind expressions to identifiers, giving calculations a specific purpose. For example, you would write the following as a formula to calculate `distance`:

```
Distance = Velocity * Time
```

As the values of `velocity` or `time` change, `distance` is automatically updated. This formula-driven approach is what makes Power Fx a unique and powerful language.

Using the `Round` function, we will get a number rounded to the right of the decimal separator:

```
Round( Distance, 1 )
```

It is an always-live environment, where changes and inputs are reflected instantly, much like an Excel worksheet. With Power Fx, you'll experience that dynamic world as it uses an incremental compiler to continuously synchronize your program with the data it operates on. This real-time approach allows you to see the impact of your changes immediately, making app development more interactive and engaging. Power Fx offers a powerful formula editing interface that includes IntelliSense, auto-complete, and type verification. The approach is low-code, allowing business logic to be expressed in concise and powerful formulas. And as it is open source, it can be extended and adopted by other platforms. Moreover, Power Fx is highly versatile, catering to various development approaches, from no-code to pro-code. No-code tools in Power Apps enable you to create customizations and logic through simple switches and UI builders, while the formula bar remains accessible for more advanced editing. For professional developers, Power Fx supports formula-based components for sharing and reuse, and its compatibility with professional tools such as **Visual Studio Code** ensures a smooth transition between low-code and pro-code environments.

Power Fx principles

Power Fx is a general-purpose, statically typed, functional, and declarative computer language that follows Excel's formula language closely, offering the same structure, data types, operators, and functions. It uses a spreadsheet-like data flow engine. Formulas automatically re-calc when a dependency changes. It adds imperative logic for button clicks, state variables, and writing back to databases, employing an incremental compiler that uses status type and data flow analysis for real-time feedback to the maker. But what does this mean for you as a maker? We will describe each of the main principles shown in *Figure 3.1* with examples:

Figure 3.1 – Power Fx principles

Let's begin with the first one, general-purpose.

General-purpose and simple

Power Fx is designed with *simplicity* and *accessibility* in mind. It minimizes the learning curve by building on the knowledge you already have and keeping essential concepts to a minimum. This simplicity also benefits experienced developers by reducing the time needed to create solutions. Pure functions without side effects are preferred as they lead to more comprehensible code and allow the compiler to optimize your app effectively.

Take a look at the following Power Fx formula:

```
"Score " & Lookup( DataSource, Points > MinScore ).Points
```

It takes care of an asynchronous process (so that the UI responds smoothly), only retrieves what is needed (delegating the filter value), performs automatic type checking, and when it is running in Power Apps, the formula bar in the Power Apps editor will try to repair and propose fixes to your formula when you make an error.

Excel's formula language mindset

Power Fx embraces **Excel's formula language** to ensure a seamless transition for users familiar with Excel. It prioritizes consistency in types, operators, and function semantics to provide a smooth learning experience. Power Fx can handle many data operations, but there are certain operations or data manipulations that Excel's formula language can't handle. In such cases, **Structured Query Language (SQL)** is used since it's a specialized language that's designed for managing and manipulating relational data.

Let's look at the best practices and capabilities from Excel in Power Fx:

```
App.Formulas =
Distance = Velocity * Time;
Velocity = Value( txtVelocity.Text );
Time = sldHours.Value;
```

In the preceding example, named formulas from Power Fx are used. They improve the app's performance as they are calculated when needed and are always up to date, among providing more benefits.

Flexibility and extensibility

Power Fx encourages combining functions and features to create more powerful, versatile components. This approach simplifies code and improves maintainability. It acknowledges the need for language evolution, incorporating a version stamp in every saved document to ensure backward compatibility. This allows the language to grow and adapt without breaking existing code, ensuring a seamless transition for users.

The latest version of Power Fx is available at `https://github.com/microsoft/Power-Fx` and included over time in Power Platform, specifically in Power Apps.

You can run the Power Fx code inside a console. For that purpose, you have an executable available called **ConsoleREPL.exe** in the repository: `https://github.com/PacktPublishing/Automate-Testing-for-Power-Apps/tree/main/chapter-03`. *Figure 3.2* shows a sample from the repository running the executable and code, `sample04.fx`, where a table is created and items are added to a collection outside Power Apps:

```
Microsoft Power Fx Console Formula REPL, Version 0.3.0.0
Experimental features enabled: TableSyntaxDoesntWrapRecords ConsistentOneColumnTableResult DisableRowScopeDisambiguation
Syntax SupportColumnNamesAsIdentifiers StronglyTypedBuiltinEnums RestrictedIsEmptyArguments FirstLastNRequiresSecondArgu
ments PowerFxV1CompatibilityRules
Enter Excel formulas.  Use "Help()" for details.

> Set( Students; Table( { Name: "Judith"; Age: 15; Points: 8 }; { Name: "Eva"; Age: 47; Points: 9 }; { Name: "Enrique";
Age: 45; Points: 7 }; { Name: "Pilar"; Age: 52; Points: 10 } ) )
Students:
  Age      Name    Points
 =====  ========  ========
   15    Judith       8
   47      Eva         9
   45    Enrique       7
   52     Pilar       10

> Collect( Students; { Name: "Ponciano"; Age: 80; Points: 9} )
{Age:80, Name:"Ponciano", Points:9}

> Collect( Students; { Name: "Bruno"; Age: 17; Points: 10} )
{Age:17, Name:"Bruno", Points:10}

> Students

  Age      Name    Points
 =====  ==========  ========
   15    Judith       8
   47      Eva         9
   45    Enrique       7
   52     Pilar       10
   80    Ponciano      9
   17     Bruno       10

> Filter( Students; Age < 20 )

  Age     Name    Points
 =====  ========  ========
   15    Judith       8
   17     Bruno       10
```

Figure 3.2 – Power FX managing a collection in a console

Running `ConsoleREPL.exe`, you have an interactive console where code is executed automatically and, due to the capabilities of Power Fx, all variables are updated automatically. You can type the code examples and check the results or create a blank Power Apps app and add it in a label text value. You can use `sample03.fx` from the repository and run it to validate the behavior.

Alternatively, you can make use of the new capability to embed expressions in a string instead of splicing them together by using the & operator or concatenate function:

```
"Your best score is " & score & ", " & FirstName
```

To embed those references directly, run the following code:

```
$"Your best score is {score}, {FirstName}"
```

For better reading and integration with other products in Power Platform and beyond, we recommend that you use the latter. It will reduce the number of errors and the code will be easier to read and check.

Declarative and functional

With Power Fx, you tell your app what you want it to do rather than how to do it. This approach enables the compiler to optimize your code by performing operations in parallel and minimizing resource usage. Power Fx automatically determines types based on usage, eliminating the need for explicit declarations. Conflicting type usage will result in a compile-time error while catching errors early by determining the type of all values at compile time.

The following is an example of error handling, which uses the upcoming feature for formula-level error management to create custom errors to improve your app usability and gather error information that's useful for the user:

```
IfError( Collect( Students, { Name: txtInputName.Text } ),
    Notify("Invalid data provided. Please try again") )
```

By adhering to these design principles, Power Fx positions itself as an accessible, user-friendly, and powerful language for app development. Its simplicity, familiarity, and adaptability make it an ideal choice for new Power Apps developers looking to create engaging and versatile applications.

The flexibility and openness of Power Fx allow it to be used through every service in Power Platform, but these aspects also allow it to be used outside. We will elaborate more on this in the next section.

Power Fx architecture

Power Fx, being an open source project, opens a wide variety of scenarios. For example, **Test Engine**, an open platform for automated testing in Power Apps, is built on top of **PlayWright** (a multiplatform framework for web testing and automation, located at `https://github.com/microsoft/playwright`) and **Power Fx**, located at `https://github.com/microsoft/power-fx`.

The architecture of Power Fx allows it to run on other scenarios such as the one shared in this book's GitHub repository. *Figure 3.3* shows one of the previous samples in a console, specifically the `sample02.fx` example, where you can make a **LookUp** of the first player with more than **MinScore**:

```
Microsoft Power Fx Console Formula REPL, Version 0.3.0.0
Experimental features enabled: TableSyntaxDoesntWrapRecords ConsistentOneColumnTableResult DisableRowScopeDisambiguation
Syntax SupportColumnNamesAsIdentifiers StronglyTypedBuiltinEnums RestrictedIsEmptyArguments FirstLastNRequiresSecondArgu
ments PowerFxV1CompatibilityRules
Enter Excel formulas.  Use "Help()" for details.

> Set( Students; Table( { Name: "Nick"; Age: 21; Points: 8 }; { Name: "Lisa"; Age: 24; Points: 9 }; { Name: "Sam"; Age:
19; Points: 7 }; { Name: "Emma"; Age: 25; Points: 10 } ) )
Students:
  Age   Name   Points
 =====  ====== ========
   21    Nick       8
   24    Lisa       9
   19     Sam       7
   25    Emma      10

> Set( MinScore; 8 )
MinScore: 8

> "Score " & LookUp( Students; Points > MinScore ).Points
"Score 9"

> |
```

Figure 3.3 – PowerFx running in a console

Power Fx's heart lies in a dynamic compiler that accelerates your development experience by providing auto-complete and IntelliSense capabilities. These features enable you to identify errors in real time, notably streamlining the debugging process.

Power Fx's tooling offers no-code options for managing control formatting and setting values visible in generated code. This flexibility allows you to edit code outside the Power Apps environment without affecting the controls. A dynamic link bridges the gap between the no-code environment and your code, enabling you to view any changes upon reloading. As a reference, *Figure 3.4* summarizes the different elements of the Power Fx implementation architecture open sourced, and which are the different areas associated with the language. In the shared GitHub repository, you can play with the open sourced elements (in light color in *Figure 3.4*). From left to right, different host applications such as a console app run Power Fx applications or a web app. In the latter, the formula bar you are used to in Power Apps Studio is used with the open source Monaco editor. This interacts with `PowerFX.Core` for parsing, binding, and compiling to **intermediate language** (IL) so that it is executed in the app:

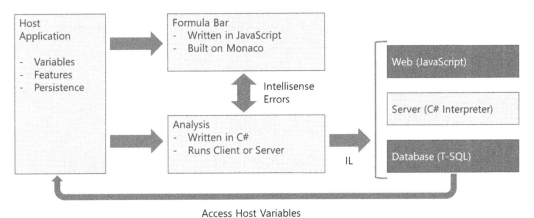

Figure 3.4 – Power Fx logical architecture

Power Fx can serve as an export format, allowing you to transfer code from visual tooling to other platforms through YAML. Power Fx exports code as YAML so that it can combine configuration data and declarative low-code statements, and it serves as the standard for other hosts (for example, Test Engine). We will take a look at YAML next.

What is YAML and why is it important?

Before diving into the definition of **YAML**, let's summarize its importance. First, YAML is a de facto standard for writing configuration files for automation in **Azure DevOps**, **GitHub actions**, or third-party services; second, **Power Fx** as an open source, low-code, general-purpose programming language that uses YAML when the host platform is not Power Platform – for example, **Test Engine for Power Apps**.

Power Fx is based on spreadsheet-like formulas. When used in Power Apps, the host platform provides the name-to-expression binding for the formulas, but when used in other hosts, YAML is the language for that binding.

Test Engine is an automated testing platform for Power Apps. It uses Power Fx test authoring, where makers can author tests in YAML format using the Power Fx language. So, as a maker developer, you will create YAML files to run tests with Test Engine, and if you use **Test Studio for Power Apps**, you could download your recorded test in YAML format, knowing that YAML will help you read, tweak, and fix your tests.

YAML definition

Welcome to the world of YAML, where you will understand the language, master its principles, and, most importantly, start using it in your apps and code testing.

Depending on whom you talk to, **YAML** stands for **Yet Another Markup Language** or **YAML Ain't Markup Language** (a recursive acronym), which emphasizes that YAML is for data, not documents. It is a human-readable data serialization format that is designed to be easy for both humans and machines to understand. It's used for configuration files, data exchange between languages with different data structures, and as an alternative to XML and JSON.

But what makes YAML so special? The answer lies in its simplicity and readability. YAML uses minimal punctuation and relies on indentation to represent the structure of data. This makes it more intuitive to read and write, especially for newcomers.

Let's look at an example YAML document to get a feel for the language:

```
# This is a comment in YAML
person:
name: Bruno
age: 17
```

```
skills:
        - Python
        - Power Apps
    - YAML
```

In the preceding example, you'll notice a few key characteristics of YAML:

- **Comments**: We start with a comment, indicated by the # symbol. Everything after this symbol on the same line is ignored by the parser.

- **Keys and values**: YAML uses key-value pairs to represent data. Keys and values are separated by a colon followed by a space – for example, name: Bruno.

- **Indentation**: The structure of data is defined by the indentation. In this case, name, age, and skills are nested under person.

- **Lists**: Lists are represented using hyphens followed by a space, as seen in the skills list.

Keep in mind that Power Apps supports a restricted subset of YAML, and this section will cover the essential constructs you need to know about.

Now that you've gotten a taste of YAML, let's talk about how you can use it in your Power Apps code testing by mixing it with Power Fx declarative low-code statements, as well as expanding the characteristics outlined.

Understanding YAML rules

Understanding the rules to create and format YAML files will help you become familiar with their structure so that you can read them easily. Let's review them in detail.

Leading equal sign – the cornerstone of expressions

First things first, every expression must start with a leading equals sign, (=). This simple yet powerful symbol ensures the following:

- Consistency with Excel uses a leading = to bind expressions to cells

- Escaping the formula language's syntax so YAML doesn't attempt to parse it

- Differentiation between Power Apps formula expressions and static scalar values

Single-line formulas – simple and effective

For single-line formulas, follow this format:

```
Name : SPACE = Expression
```

Remember to include the space between the colon and the equals sign to be YAML-compliant. The equals sign allows the rest of the line to be treated as Power Fx.

Multiline formulas – expanding your expressions

For more complex expressions spanning multiple lines, use YAML's block scalar indicators:

```
Name : SPACE ( | or |+ or |- )  = Expression-Line  Expression-Line  ...
```

Ensure all lines in the block are indented at least one space from the first line's level.

In this extract of a test for a Power Apps Training app, you can see an example of this:

```
...
  testCases:
    - testCaseName: PowerAppsPATApp
      testCaseDescription: 'Onboarding steps of user'
      testSteps: |
        =Select('ButtonNext');
        SetProperty('Toggle1'.Value, true);
        Select('Toggle1');
        Select('ButtonNext_1');
...
```

In this example, you can see a Test Engine or Test Studio test case in YAML format and PowerFX operators and functions as `Expression-Line`.

Component instance – bringing components to life

To instantiate components, use YAML object notation. The object's type is established with the `As` operator, while container controls can be nested:

```
Name  As  Component-Type  [ .  Component-Template ]  :  ( Single-
LineFormula or Multi-Line-Formula or Object-instance )  ...
```

Indent all lines in the block by at least one space from the first line's level.

Here, `Component-Type` can be any canvas component or control, and `Component-Template` is an optional specifier for components that have templates.

The following is an example where a gallery is used:

```
"'A name with a space' As Gallery.verticalGallery":
    Fill: = Color.White
    Label1 As Label:
        Text: ="Hi Judith"
        X: =15
```

```
        Y: =47
        Fill: |
            =If( Upper( Left( Self.Text, 6 ) ) = "error:",
                Color.Red,
                Color.Black
            )
```

You can use the Power Apps Source File Pack and Unpack utility to manage source code for your Power Apps in YAML. Check the GitHub repository at `https://github.com/microsoft/PowerApps-Language-Tooling/`, where the previous gallery is an example that presents a component and its elements.

YAML compatibility – harmonizing Power Fx and YAML

To ensure a smooth integration, be aware of YAML compatibility. For example, comments in YAML and Power Fx have different delimiters. Use `//` or `/*` and `*/` for comments within formulas.

Also, some Power Fx and YAML grammar aspects might clash, causing errors. For example, using a number sign, #, within a text string or identifier name in a single-line formula could lead to errors. To avoid this, use multiline formulas instead.

Record YAML allows the same name map to be reused, with b: silently overriding the a: definition. As an example, these cases will throw an error:

```
Text: ="Hello #Enrique"
Record: ={ a: 33, b: 7 }
```

We can fix both issues by writing the code as follows:

```
Text: |
="Hello #Enrique"
Record: |
A: =33
B: =7
```

By understanding these key principles and constructs, you're well on your way to harnessing the full potential of Power Fx and YAML in your apps. In *Chapter 5*, you will understand the importance of Power Fx when creating test cases, while in *Chapters 6* and *7*, you will see how Power Fx is used in Test Engine using YAML files.

Let's finish this section with a full example of a test suite and the test cases of a calculator app in YAML format:

```
testSuite:
  testSuiteName: Calculator
  testSuiteDescription: Verifies the calculator app.
```

```
persona: User1
appLogicalName: new_calculator_a3613
testCases:
  - testCaseName: Default Check
    testSteps: |
      = Assert(Calc_1.Number1Label.Text = "100", "msg1");
        Assert(Calc_1.Number2Label.Text = "100", "msg2");
  - testCaseName: Check subtraction
    testSteps: |
      = Select(Calc_1.Subtract);
        Assert(Calc_1.ResultLabel.Text = "0", "msg3");
        Screenshot("Calc_end.png");
testSettings:
  locale: "en-US"
  recordVideo: true
  browserConfigurations:
    - browser: Chromium
environmentVariables:
  users:
    - personaName: User1
      emailKey: user1Email
      passwordKey: user1Password
```

Check out the different rules that are applied and how the definition matches the test suite. Take this chapter as a reference and use the online documentation to run the examples with Power Fx, specifically the functions highlighted in the next section.

Exploring the top test functions in Power Fx

You can use any function available in Power Fx, but there are some top functions you will use in every test. We'll dive into six important functions: `Select`, `SetProperty`, `Wait`, `Assert`, `Screenshot`, and `Trace`. By the end of this section, you'll have a firm understanding of how these functions work and how to use them in your apps and code testing.

As we discussed in *Chapter 1*, a large portion of the testing in your low-code apps will be UI tests, where you reproduce the actions you expect your users will follow. To validate the expected experience and results of your users, you need to follow three steps:

1. Execute the steps to get and set the data in your app that confirm a defined use case. For example, in a weather app, click on a text box, enter the city's location, and click on the respective button to get the required weather details.

2. Validate that data against the expected behavior. In the weather app, it would validate that the label shows the weather from the city.

3. And, optionally, log additional information you would like to include as part of the context of the test.

Let's dive deeper into each step and function.

Simulating user interactions

The `Select` function in Power Apps simulates a user selecting an action on a control, which triggers the `OnSelect` formula on the target control to be evaluated. This function can be used to propagate a select action to parental control or to perform different actions for different controls within a gallery. The `Select` function can be used with controls that have an `OnSelect` property and can be used in behavior formulas. However, it cannot be used across screens, nor can a control select itself directly or indirectly through other controls. The `Select` function can also be used with galleries to specify the row or column to select and the control within that row or column to be selected. Here's the `Select` function's syntax:

```
Select( Control, Row or Column, Child Control )
```

Here, `Control` is the control property you want to select on behalf of the user. `Row or column` and `Child control` are not required. You could use them to select a specific element of a gallery (the number of rows or columns) and to identify a child control of the previous control property.

With the `SetProperty` function, you can simulate user interactions in Power Apps Test Studio, as if the user entered or set a value on a specific control. This function is exclusive to the Test Studio environment. Here's the `SetProperty` function's syntax:

```
SetProperty(Control Property, Value)
```

Here `Control Property` is the control property you want to set on behalf of the user, and `Value` is the desired property value.

Several examples are shown in *Table 3.1*:

Control	Property	Example expression
TextInput	Text	SetProperty(txtLocation.Text, "Madrid")
RichTextEditor	HtmlText	SetProperty(rteForecast.HtmlText, "<p>Heavy Rain</p>")
Checkbox	Value	SetProperty(chbTemperature.Value, false)
Rating	Value	SetProperty(rtnFeedback.Value, 5)
DatePicker	SelectedDate	SetProperty(dteDateRange.SelectedDate, Date(2023,5,10))
Radio	Selected	SetProperty(radCelsius.Selected, "Yes")
Dropdown	Selected	SetProperty(drpSkyMode.Selected, {Value:"Cloudy"})

Table 3.1 – Examples of SetProperty usage

The `Wait` function will let you wait for the property of a control to have the expected value:

```
Wait( Control, Property, Value)
```

Here, `Control` is the control you want to wait for on behalf of the user, `Property` is the property of the control to look for, and `Value` is the desired property value until the wait is over. The following is an example of waiting until the text for controlling weather presents the expected result:

```
Wait(lblWeather, "Text", "24")
```

There are many other functions that you could use. Check out the examples in the Test Studio documentation and the list of supported control properties and expressions.

The true-or-false test

An *assertion* is a crucial part of any test as it checks whether a specific condition is true or false. If the condition evaluates to false, the test fails. Assertions help ensure that your app is functioning as intended by validating various aspects such as label values, list box selections, and other control properties. Here's an example of the `Assert` function's syntax:

```
Assert(expression, message)
```

Here, `expression` is the condition that evaluates to true or false, and `message` is an optional description of the assertion failure.

As an example, we have a simple calculator app with two labels for number input, one label for calculated results, and four buttons for **Add**, **Subtract**, **Multiply**, and **Divide**. The following code will ensure that the app works as expected when we enter two different numbers, 25 and 35, click the **Add** button, and get the expected result of 60:

```
SetProperty(txtNumber_1.Text, "25");
SetProperty(txtNumber_2.Text, "35");
Select(btnAdd);
Assert(lblResult.Text = "60", "Confirm addition result");
```

In *Chapters 5*, *6*, and *7*, you will learn more about Test Studio and Test Engine and drill down into the usage of these functions in some real examples.

A peek behind the curtain

Sometimes, it's not immediately clear what's happening within your app. Enter the `Trace` function, a powerful diagnostic tool that captures behind-the-scenes information, creating a timeline of app events. The output from `Trace` appears in the **Power Apps Monitor** tool, and if you've configured it, in **Azure Application Insights** as well.

Here's the syntax for the `Trace` function:

```
Trace( Message [, TraceSeverity [, CustomRecord [, TraceOptions ] ] ]
)
```

Here, `Message` is the information to be traced, `TraceSeverity` is an optional severity level, `CustomRecord` is an optional record containing custom data, and `TraceOptions` is an optional setting to ignore unsupported data types.

In *Chapter 2*, we presented an example of the `Trace` function. We walked you through creating a button in **Power Apps Studio** that uses the `Trace` function, and how to view the results in **Power Apps Monitor**. As you saw, it allows you to add any additional context to better understand the flow of your test and code, adding custom logging. In this section, we will add additional information to our calculator action:

```
Trace("CalculatorAddition",
TraceSeverity.Information,
{
AppName: "Simple Calc",
UserName: User().FullName,
Screen:"CalcScreen"
}
)
```

Apart from custom logging, you may need to see what the app looks like. The `Screenshot` function will let you capture a screenshot of the app or component at the current point in time of the test. The output will be saved as JPEG or PNG files and to the output folder provided:

```
Screenshot( Filename )
```

Here, `Filename` is the name with the extension of the screenshot to be taken. Here's an example that was previously presented in the test suite for the calculator app:

```
Screenshot("Calc_end.png");
```

With these functions, you'll be well equipped in your testing journey of using **Power Fx**. Keep practicing, and you'll soon master the art of using them to create robust apps. On the other hand, **artificial intelligence** (**AI**), specifically **generative AI**, is taking the world by storm and simplifying development and testing thanks to its natural language processing capabilities. In the next section, we will present how it is adopted in Power Apps and provide an example to show how it can support Power Apps test files with Copilot.

Natural language to Power Fx

Part of the dream of low-code platforms is to democratize development so that everyone without computer science knowledge can develop their tools or apps. In that vision, a way to interact with a computer in your terms means using your voice or your writing is the most accessible way. Recently, the performance and capabilities of AI models, such as those from **OpenAI** (**GPT**) and, most recently, **ChatGPT**, have accelerated its adoption in many areas, and Microsoft products and services, specifically Power Platform, have not been an exception.

This integration in the platform accelerates the development process, and it will soon impact the way we test or design our apps. That's why we want to share some of the AI features in the platform, and specifically, what is happening in the ecosystem that could bring value to the low-code application life cycle.

AI and Power Platform

Figure 3.5 shows some of the latest features (at the time of publication) of using AI in Power Platform. Specifically, it specifies how Power Apps is leveraging **Azure OpenAI GPT-3** and the **PROSE** SDK. PROSE is a Microsoft Research team that assists programmers through automatic program repair and conversations AI; you can learn more about it at `https://www.microsoft.com/en-us/research/group/prose/`. For example, one of its capabilities is **programming by examples** (**PBE**), which enables users to create code from input-output examples, or instead of starting an app from a template, create a working sample with an image of your app. Finally, GPT-3 allows you to generate Power Fx queries in natural language, taking into account the context of your app to generate one or a few of the most relevant formulas for you to select from:

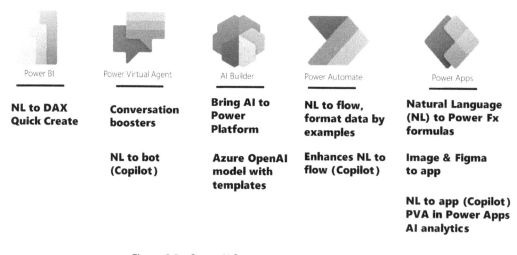

Figure 3.5 – Some AI features across Power Platform

A common use case is to generate the code required to generate an output based on the available input data. Imagine a table where you want a name in a specific format and give an example input. PBE allows you to generate the Power Fx formulas through the Power Apps **Ideas** feature, as presented in *Figure 3.6*:

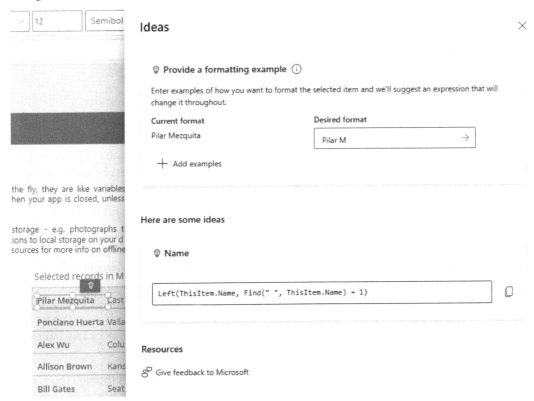

Figure 3.6 – Power Apps Ideas in action suggesting formatting code

As you can see in the following code associated with the preceding figure, there is a table with people's names and locations (Data), and we want to format it with the name and the first letter of the surname. We bring that as an example (Input), and based on the data available, Ideas presents some to apply that format (AI-Generated Output):

```
Data: Pilar Mezquita
Input: Pilar M
AI-Generated Output: Left(ThisItem.Name, Find(" ", ThisItem.Name) + 1)
```

Power Apps remains steadfast in upholding Microsoft's responsible AI principles, which prioritize fairness, inclusiveness, safety, reliability, and the safeguarding of privacy and security. By integrating **GPT-3** and **PROSE** into Power Apps, extensive training and built-in safety controls have been employed to prevent the generation of harmful outputs. In the spirit of transparency and openness,

Power Apps is dedicated to offering clear and succinct information about its AI-driven features and capabilities. Microsoft is devoted to empowering users with the knowledge and control necessary to comprehend and manage their interactions with AI. We encourage you to check out the latest updates from the Power Apps blog at `https://powerapps.microsoft.com/en-us/blog/` and the responsible AI principles at `https://www.microsoft.com/en-us/ai/responsible-ai`.

One of the latest announcements is related to the concept of **Copilot**. Let's take a closer look before we close this chapter.

Your Copilot

In 2021, GitHub announced a new service called GitHub Copilot. It was a new cloud-based AI tool developed between GitHub and OpenAI to help users auto-complete the code of their apps in any potential language. It is integrated into the development environment being used, such as Visual Code, and it assists the user in coding, commenting, or explaining the code itself. It starts by using a piece of text in natural language from the user and a prompt, which specifies how to generate code, rewrite it, or comment on it. You can learn more at `https://github.com/features/copilot`.

Microsoft has extended this concept to all its services, including Microsoft 365 Copilot, Power Platform Copilot, Dynamics 365 Copilot, and more. Specifically, Copilot has been added to Power Apps as an AI assistant in many ways: from starting from a prompt description of the app to describing the type of data so that Power Apps can create a working draft app. It could benefit the final user of the app so that you will be able to add natural language capabilities to your apps. You can find more details at `https://powerapps.microsoft.com/en-us/blog/announcing-a-next-generation-ai-copilot-in-microsoft-power-apps-that-will-transform-low-code-development/`.

We will close this chapter by reinforcing the importance of the notion of Copilot through the peer programmer in Power Apps Studio, as well as how it can act as a potential assistant to help you better understand your code. As a final example, *Figure 3.7* presents a Power Apps Test Studio exported file opened with GitHub Copilot Labs. All the comments from the YAML file have been documented by the *List Steps* feature:

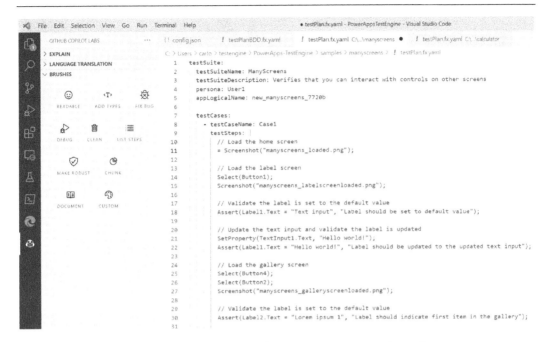

Figure 3.7 – Test Studio YAML file commented by GitHub Copilot

As GitHub Copilot explores test generation in JavaScript and TypeScript, the whole ecosystem extends in terms of its code automation capabilities. This will accelerate in the coming years, and it will help it get closer to low-code vision development.

Summary

In this chapter, we introduced Power Fx, the low-code language for Microsoft Power Platform, and focused on its applications in testing and enhancing low-code productivity. We covered key topics, including the relationship between Excel and Power Fx, understanding the role of YAML in formatting Power Fx, how to use Power Fx to test apps, and discovering new AI-driven capabilities for natural language development in Power Fx.

Upon completing this chapter, you gained the ability to navigate the Power Fx language, comprehend YAML formatting for Power Apps, and effectively apply test functions. Additionally, you learned about the platform's natural language capabilities. Armed with these skills, you are now better equipped to enhance your Power Apps development and testing processes. In the next chapter, we will review some of the concepts mentioned in the previous chapter and cover various Power Apps capabilities that will allow you to better plan an app's development and its testing phase.

4

Planning Testing in a Power Apps Project

In the previous chapter, we ventured into the realm of Power Fx, the low-code language that powers Microsoft Power Platform, diving into its application in testing and low-code productivity enhancement. We navigated the transition from Excel examples to Power Fx, understanding YAML's role in Power Fx formatting, and utilizing the essential test functions from the Power Fx reference. We also saw the emergence of natural language capabilities in adding Power Fx code and programming by example.

In this chapter, we're shifting gears to focus on the application of testing within the scope of Power Apps development, tailored to your organization's specific maturity level in testing. We'll kick things off by exploring how different stages of organizational maturity impact your approach to testing in Power Apps. We'll discuss an end-to-end example, highlighting tools that are particularly pertinent for different roles in Power Apps projects. Then, we'll demonstrate how effective test planning can be enhanced by incorporating user feedback and performance metrics, methodologies that were covered in previous chapters. Our map for this exploration is as follows:

- **Different levels of maturity around testing in your organization**: We'll introduce you to the testing process when building with Power Apps while considering your organization's maturity model and role

- **From citizen developer to fusion team**: We'll describe the different tools that apply based on your role and activity, and how this ties into the broader view of a fusion team and Power Apps projects

- **How test preparation benefits your user feedback**: We'll delve into how using tools and processes introduced in previous chapters, such as feedback collection and monitoring, can be connected to the definition of test cases and scenarios

By the end of this chapter, you will have gained an understanding of how to apply previously learned testing concepts to a Power Apps development project and understand the maturity of adopting testing processes or tools. You will have a clearer picture of the importance of testing in managing the evolution of your app and how defining your app helps in planning related testing activities. Let's delve deeper into these fascinating topics.

Technical requirements

To test and follow some of the capabilities of this chapter, you will need a few technical requirements.

All material will only require the following:

- A stable internet connection and a compatible browser, such as Google Chrome or Microsoft Edge.

- You will also be required to access Power Apps and other online resources, such as GitHub.

- We will use a calculator app named Power Calc. Guidance and samples are located at `https://github.com/PacktPublishing/Automate-Testing-for-Power-Apps/tree/main/chapter-04`.

> **Important note**
>
> We recommend that you create a Power Apps Developer Plan so that you can test all functionality moving forward. You can follow the instructions at `https://powerapps.microsoft.com/en-us/developerplan/`, where you will get a free development environment to develop and test apps.

Planning based on your organization's maturity

Some believe that change management doesn't connect to an organization's maturity, but in reality, they're deeply linked. In today's fast-changing world, leadership is more than just setting a vision – it's about driving alignment and facilitating change. With flatter structures and a sharper focus on customers, organizations demand more from their employees. You cannot just throw technology into an organization and expect it to be used right away. A change management process for the adoption of an agile process and testing mindset is critical. Measuring adoption maturity in an organization involves evaluating the extent to which new processes, tools, or technologies have been integrated and are actively used, as well as the level of proficiency and alignment with business objectives.

We've already been talking about the importance of testing, the business impact, the support, and the tools needed. We presented **Power Platform's adoption maturity model** in *Chapter 1, Software Quality and Types of Testing*, as a guide to adopting a low-code culture to test apps in your organization, and we described SDLC and ALM as the vehicle where you can run testing activities. Indeed, the title of this chapter is derived from the Power Apps guidance documentation, which can be found at `https://learn.microsoft.com/en-us/power-apps/guidance/planning/introduction`, where you can review a summary of activities for *Planning a Power Apps project*, connected with the topics of *Chapter 1*. If you review the documented planning, it is assumed everything is set up. The organization has everything in place, but the reality is a bit different.

With the maturity model in mind, and the steps required to better adopt any technology change in an organization, we will describe which tools and processes you should use for a better testing process. The answer is connected to how any change is adopted personally and in any organization, and we want you to think about where you and that organization are positioned to best include any testing activity.

One of the most recognized firms in change management and adoption is **PROSCI**, which has a model you can follow to guide a technology adoption called **ADKAR**. The acronym stands for Awareness, Desire, Knowledge, Ability, and Reinforcement – each representing a crucial step in the change process:

1. **Awareness**: This initial stage is where the need for change is recognized. It includes understanding the nature of the change, why it is needed, and the risks of not changing. Effective communication is key at this stage to create a sense of urgency and kickstart the change initiative. For our objective, this translates to *Why should I test my apps?*

2. **Desire**: At this stage, individuals within the organization acknowledge the benefits of the proposed change and develop a personal motivation to support and participate in it. Desire is influenced by factors such as the perceived benefits of change, organizational culture, and individual temperament. For our objective, this translates to *It seems too much work, the organization should include it later; what's in it for me if we implement automated testing?*

3. **Knowledge**: Here, individuals learn about change and what it requires from them. This stage involves acquiring the skills, abilities, and understanding necessary to implement the change. Education and learning initiatives are vital at this stage. For our objective, this translates to *How do I execute automated tests in Power Apps?*

4. **Ability**: This stage focuses on the practical application of the acquired knowledge. Individuals should now have the capability to demonstrate the new skills and behaviors that the change requires. It might involve changes to processes, systems, job roles, and more. For our objective, this translates to *Can I efficiently run and manage automated tests now?*

5. **Reinforcement**: The final stage of the ADKAR model is where the change is reinforced to make it stick. This involves providing ongoing support, celebrating successes, and embedding the changes into the organizational culture to prevent fallbacks to old habits. For our objective, this translates to *How do we ensure automated testing becomes a standard practice?*

Let's mix these elements into five examples to help you understand what you should consider when planning testing for your app. This is not a hard differentiation but guidance through examples to help you understand the changes or states you need to consider to fully embrace every platform capability or process activity.

Example 1 – testing during the early stages of software development

As an organization embarks on its journey to adopt Power Apps, it is crucial to plan testing, even at the nascent stages of development. At this stage, you should focus on functional testing and exploratory testing. These forms of testing are crucial in determining the basic functionality of the app and discovering any potential bugs early on. The Power Platform adoption maturity model's Initial phase coincides here, where an organization begins experimenting with Power Apps capabilities. Here, you usually test manually for apps as you develop them.

Example 2 – integrating automated testing

As the development maturity progresses, automated testing should be introduced. You won't be able to scale the complexity of your app and make the testing at organization-level practice without automation. The Power Platform adoption maturity model's Repeatable phase corresponds with this step, suggesting that the organization is now creating apps consistently and recognizes the benefit of automated testing to handle repetitive, standardized test cases. Here, you will consider regression and integration testing, as well as start using Power Apps Test Studio by adding some tests and running them manually or semiautomated with the benefits we have been highlighting.

Example 3 – ALM evolution – performance and usage

Once the organization starts to embrace the new processes and tools, the importance of performance and security testing becomes evident. The organization's Power Platform adoption maturity stage now becomes *Defined*, where applications are developed and deployed systematically. These tests ensure the app runs smoothly under load and is secure from potential breaches. At this point, the importance of monitoring is important for its performance and usage understanding. The availability of ALM scenarios needs to consider adoption activities for multidisciplinary teams with different levels of tech knowledge, but that participate in the development and testing process.

Example 4 – UAT and business insights

When the organization reaches the Managed stage in the Power Platform adoption maturity model, **user acceptance testing** (**UAT**) should be performed. It suggests a well-established process for creating apps. UAT allows users to interact with the app, ensuring it meets their expectations and requirements. Training should be provided to make sure users can efficiently utilize the environment through clear communication and governance in place. UAT is not just about validating the app functionality, but the business expectation. You can measure and connect both perspectives. As we mentioned earlier, there could be different degrees of maturity in every example, so don't feel you should be doing all or nothing. At this point, automation and the evolution to connect business metrics and insights from the data and tests is a clear benefit but a high maturity signal.

Example 5 – full automation and regular quality checks

Finally, where the change is supported and sustained, the organization should be in the *Optimizing* phase of the Power Platform adoption maturity model. Regular quality checks should be implemented to ensure the continual functionality and performance of the app. The feedback loop from the users should be well established, and regression testing should be scheduled whenever updates or patches are applied to the app. At this stage, the possibility of further automation allows a complete collaboration view, where actions could be embedded in multiple communication channels or tools. As we mentioned in the previous example, you can start automating to test if business goals are met.

Where are you and your organization on that journey? This will affect which tools you could use and learn, but based on your profile, be it a pro-developer or citizen developer, the right processes and training should be in place to help make it real.

In the reference documentation that we saw earlier, *Planning a Power Apps project*, there is a maxim that you should keep in mind through the whole planning: the cost of doing nothing. You need to figure out the effort you need to put into the details of your tests or their automation. Could I spend time manually doing those steps? Which activities should I include for better testing? For example, creating functional documentation and use case descriptions will help you create different tests. They could help you validate the behavior of the app so that your users are satisfied with the expected result, and they will allow you to anticipate breaking changes in previous app versions.

After reading these five examples about the implications of change management needed personally and in the organization, you might be thinking about the tools and processes to be adopted, and how the Power Platform adoption maturity model guides your organization not just through the technical aspects of developing an app but also through the associated organizational and people-oriented challenges.

Understanding this change will help you to better understand the implications of the benefits and needs of Power Apps project planning, its costs (from doing nothing to the cost of adopting a change management process, tools, and processes to bring it to life), and its benefits.

Let's review two examples in that journey, based on the reference architecture for the Power Platform application life cycle at `https://learn.microsoft.com/en-us/azure/architecture/solution-ideas/articles/azure-devops-continuous-integration-for-power-platform`.

The first one will focus on *Example 2*, which is connected to the repeatable phase of the maturity model, where you are a solo citizen developer building apps in your department and you need some processes and tools to anticipate issues happening in your apps as they grow in functionality and use. *Figure 4.1* presents the services involved in the minimal automated testing mentioned:

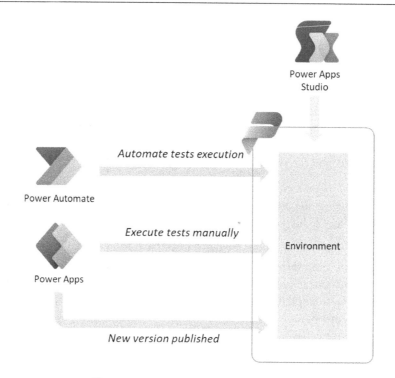

Figure 4.1 – Minimal automated testing

In this example, the reference architecture has been greatly simplified because there is no DevOps process in place. You, as a citizen developer, create and update the app's publishing in a Power Platform environment and use Power Apps Test Studio to create or record test suites and test cases that validate your app's functionality. Power Automate is used to periodically execute your tests and Power Fx allows you to add additional actions and run the Power Apps tests. Although the best practice should be to run those tests through Azure DevOps or GitHub, the maturity of your organization and the availability of those tools and their adoption directly affects the desired one.

When we investigate the second scenario, let's say *Example 4*, which is connected to the capable phase of the maturity model, where a Center of Excellence has been established, monitoring and app usage are available through app telemetry, and the customer environment is used for defined use cases. You are a developer working in the same department but, as the company has matured in its low-code adoption, the processes for low-code and pro-developers in the app are very well established. The app has expanded functionality with custom controls and additional cloud services, and you are part of the team working on this business-critical app. You will continue to use some processes and tools to anticipate issues happening in your apps and hold responsibility for the potential issues of the app in published versions, working closely with the other members of the team. *Figure 4.2* presents a more advanced but simplified version of the reference architecture, and it presents the services involved in the evolution of ALM:

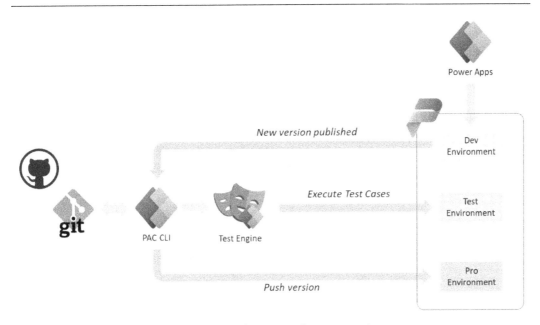

Figure 4.2 – ALM evolution – performance and usage

In this example, you, as a developer, create and update the test cases, and the necessary Power Apps functionality is assigned. There are multiple Power Platform environments, and Power Apps Test Studio is used to create tests, but the whole process is automated through GitHub Actions, which is automated with Power Platform's **command-line interface** (**CLI**) using Power Apps Test Engine. The DevOps process in GitHub is used to periodically execute your tests in the test environment and publish them in every environment defined.

It would be great to deep dive into the first scenario to better understand the tools and processes involved. In the next few chapters, you will go fully hands-on, and we will anticipate the processes involved. Keep reading!

Running a low-code developer example

As digital transformation continues to evolve and empower non-technical personnel, the role of a **citizen developer** has emerged to the forefront. In this role, you are typically a business user who creates applications intended to solve departmental or company-wide issues using low-code or no-code platforms such as Microsoft's Power Apps. However, as your apps grow in complexity and user base, there comes a need for a more structured approach to managing changes and preventing potential issues. This is where our first fictional scenario comes into play.

You are responsible for creating and managing a series of apps that are critical to the day-to-day operations of your department. Your apps have started to expand in functionality and use, making it necessary to introduce measures to anticipate and manage potential issues.

You have developed a calculator for financial math, and you've included some calculations that are used extensively in your department. You are planning to add tests to your app so that you are confident future changes will not mess up the previous calculations. You would also like to automate the whole thing so that you can receive the health of the app periodically and fix or roll back conflicting changes.

What do you need to consider when planning testing activities?

Based on the SDLC from *Chapter 1*, we will mention the related testing activities for each stage. *Figure 4.3* presents the different stages you know from the previous chapters. We will follow these different stages for our calculator:

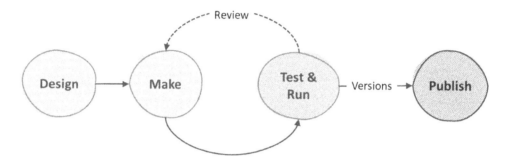

Figure 4.3 – SDLC simplified for low-code

Testing is part of the overall app development life cycle and is where app development is a critical first step. As a developer, you could rely on some processes such as **test-driven development** (**TDD**), which relies on app requirements being converted into test cases before the app is developed. However, the testing approach in Power Apps requires the app itself to define the different test cases, or in a more advanced scenario, you could develop and run in Test Engine. Is there any previous activity we should check? You started developing the calculator app, but although you considered the different requirements needed in your organization, you did not write them down. Now, you would like to add someone else so that other people can support and make changes. Moreover, to create test cases, they need to understand the app and how it works.

In the design stage, you will go through the app *requirements* to understand the app better, and with that knowledge, you can follow several recommendations:

- Follow a clear nomenclature to better understand each element involved. For example, prefixes such as *btnXXX* help you understand the different code or actions. Remember the best practice guidelines mentioned in *Chapter 1*.

- Use a collaborative place in your department to share documentation about the app, which includes its objectives and capabilities. You could use open source apps to create automatic documentation and complement your initial requirements material. Additionally, you could use the **Power Platform CoE Toolkit**, an example of which can be found at https://github.

`com/modery/PowerDocu`, to generate Word or markdown to be included in that place, such as SharePoint Site, Teams, or GitHub Wiki.

- Think about recording an app walk-through, where any user could understand some of the complex functions or screens.

In the *Make* and *Test and Run* stages, you use Power Apps Test Studio to generate different test cases, through the recording capability you will see in the next chapter, or just using your skills of Power Fx. That *definition and development* of tests should follow the requirements information and it will be part of the actual app code. Test cases are stored in the app, as you use Test Studio. This is important as you may change some naming you did not make the first time, such as changing the name from `Button01` to `btnNumberOne`, and the test cases created will change automatically. In this case, the main recommendations would be as follows:

- Find the sweet spot for testing the number of test cases and add them over time based on their usage and importance. The more complex an area becomes, the more important its test cases will be. The more important an area is for your users, the more test cases and issues you will find.

- Sometimes, you may use a custom component or custom connector to extend some capabilities, developed in-house or externally, and you may want to add specific test cases for more complex scenarios with the component. With Power Fx, you could develop code to use the component and change the values in the test or, as described later in this book, simulate the connector call. However, you may find it useful to create a canvas app not available for the users so that you can run more complex test cases for that component.

- If you were responsible for the code of the custom components, you would include additional tests and run them apart from the ones in Test Studio.

- Documentation is great, but commenting on your code, test cases, or app code and writing explanatory code is as good as the best documentation.

When it comes to the *execution and reporting* aspects of the tests, they need to be running periodically, but at least they will run when you publish the app with Power Apps Test Studio. What does this mean? You may find some issues as you publish the app. So, there are some recommendations to look into:

- As the app moves into a critical component of your department, and Power Platform's maturity adoption grows, different environments need to be in place to test before the users use it. You will use Test Studio in a development or test environment and, after checking it works as expected, move it into production.

- As part of the test suite definition, you should determine how to check for the running test results. They could be stored so that they're checked manually, sent by email, or integrated into a dashboard. You get a result file for the test's run, and with that content, you can make your own process for reporting, history, or actions. The available resources and processes in the organization will guide that decision.

- As the app and the organization grow, more complex scenarios will appear, so mixing pro-developers and citizen developers should be considered. This will affect how the test cases citizen developers create need to be added to the ALM process selected. For example, you could use Test Engine and/or the Power Platform CLI to run the tests, integrated with the files created in Test Studio.

Finally, *monitoring* will help you get a view of the running app from your users. The different tools mentioned earlier will apply here, such as Power Apps Monitor and Azure Monitor. Here are some recommendations to follow:

- Run the app with a monitor session as it will give you an under-the-hood view of potential issues, from performance to accidental errors. You could invite other users to see the application monitor data so that you can explain or ask for guidance on issues.

- Power Apps Monitor allows you to download and upload data, so you can check later, or even attach the activity to an issue you need to fix and through a test case.

- Azure Monitor, with Application Insights, will give more complete reporting, with ad hoc queries to understand usages and screen activity. For example, you could add **Traces** with Power Fx, to enrich telemetry with business metrics to review with the telemetry logs.

> **Note**
> We will focus on the actions needed to plan, execute, and automate the tests for the sample app. Explore the documentation and next few chapters to set up a full scenario, specifically with Test Studio and Test Engine.

Testing and documentation in the calculator app

In this first scenario, the following process describes the story of the app, and it points out the requirements for test automation:

1. You have created your app – in our case, Power Calc – in an environment available for the users of the department.

    Check out the example in this book's GitHub repository: `https://github.com/PacktPublishing/Automate-Testing-for-Power-Apps/tree/main/chapter-04`. Since it's a canvas app, you need to import it via canvas app import.

2. From the `PowerDocu` repository, you downloaded the `PowerDocu v1.2.4-self contained` ZIP file and used the exported calculator file from this book's repository to enrich the app description. *Figure 4.4* presents the detailed documentation. As an example, it presents code information and flows presented in the app:

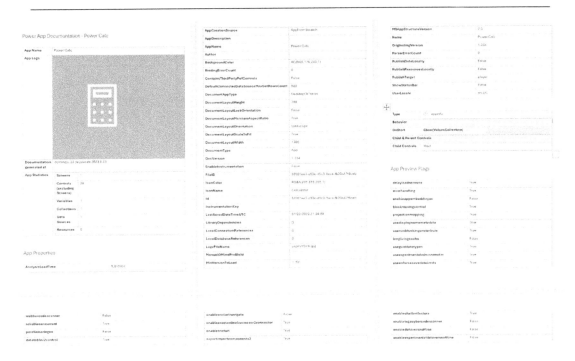

Figure 4.4 – PowerDocu export from the sample calculator app. Note
that this screenshot only intends to show the layout

This helped you better understand the app, as well as some of its inner details, such as `OnStart=Clear(ValueCollection)`, which stores the app calculations you may have forgotten to mention to the new member, but generating the documentation helps you review all areas of your app:

1. You started developing the app and added code to manage decimals, but it has introduced a bug in the app and there are errors about how operations were calculated.

2. You didn't document the application; you need to remember to periodically run the PowerDocu export, but while you start writing a help guide, you understand writing the expected use cases of the app would complement and validate the expected functionality. You use Power Apps Test Studio's **Recording mode** to create the use cases, such as the mentioned testing decimals numbers. *Figure 4.5* shows **Recording mode**:

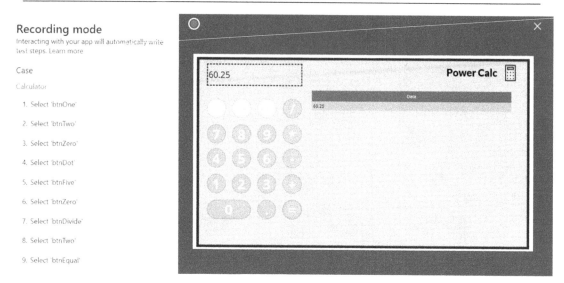

Recording mode
Interacting with your app will automatically write test steps. Learn more

Case

Calculator

1. Select 'btnOne'

2. Select 'btnTwo'

3. Select 'btnZero'

4. Select 'btnDot'

5. Select 'btnFive'

6. Select 'btnZero'

7. Select 'btnDivide'

8. Select 'btnTwo'

9. Select 'btnEqual'

Figure 4.5 – Recording mode in Power Apps Test Studio

3. Once you have created the test, you add some Power Fx code to assert the expected result and send information to your email. The following code was added to validate a correct operation:
    ```
    Assert(lblOperation.Text = "60.25","Confirm calculation with
    decimals is right Expected : 60.25 , Actual : " & lblOperation.
    Text).
    ```

This is included in the **Advanced Tools | Open Tests** option of Power Apps Editor, and then added as a step in the test, as shown in *Figure 4.6*:

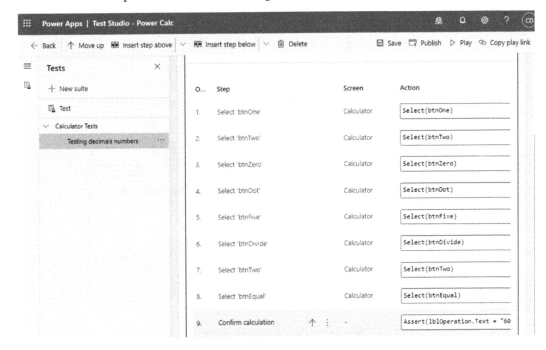

Figure 4.6 – Test case for testing decimal numbers

It is important to note that to include any service in **Power Apps Test Studio**, you need to add that service to the app itself. As shown in *Figure 4.7*, to be able to send an email, you need to do the following:

i.   Click **Add data** in your app and look for **Office365Outlook service**.

ii.   After confirmation, you should see it in the list:

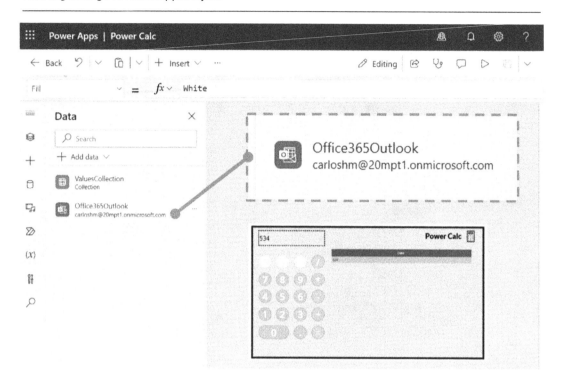

Figure 4.7 – Office365Outlook service added

The information will be sent as inline JSON content in an email, as defined in the test completion events shown in *Figure 4.7*:

Figure 4.8 – Test completion events executing Power Fx commands (for example, sending email)

4. You periodically check the execution of the tests, but usually, there are no errors in the execution, so you include a call to Power Automate to check the tests and check for errors, getting specific alert information when something goes bad.

    For this scenario, you will create a Power Automate flow that will run the test and get the results of the test file. In a more mature scenario, this will run in Azure DevOps or GitHub Actions. For this example, you follow these steps:

    i.   Navigate Power Automate in the app and create a blank flow.

    ii. Add a variable to store in the flow of the test results, parse the JSON file to detect errors, and send an email for the Test Suite complete event.

    iii. You should use the file sent previously to generate the schema – that is,
         ```
         { "EndTime": "2023-06-22T23:20:55.526Z", "StartTime":
         "2023-06-22T23:20:22.980Z", "TestSuiteDescription": "",
         "TestSuiteId": "6252d1d6-f49e-41e4-88fa-5d95f88cf66c",
         "TestSuiteName": "Calculator Tests", "TestsFailed": 1,
         "TestsPassed": 0 }:
         ```

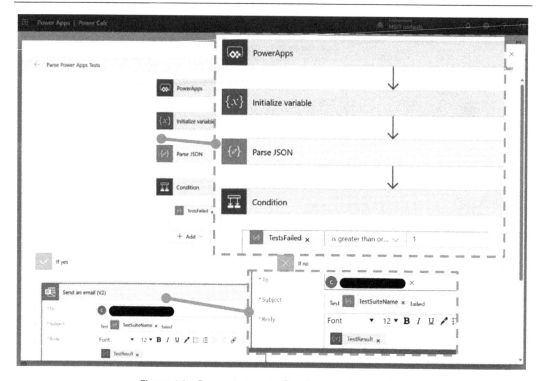

Figure 4.9 – Power Automate flow to parse test results

Figure 4.8 shows a basic workflow of how you can leverage the various Microsoft Power Platform tools to automate reporting and ensure the quality and functionality of your apps. In this case, the OnTestSuiteComplete event from *Figure 4.7* will be changed for the following code:

```
ParsePowerAppsTests.Run(JSON(TestSuiteResult, JSONFormat.
IndentFour))
```

As the organization's maturity grows, the scenario will grow in complexity and richness to deliver products and solutions more efficiently. This may involve utilizing some of the advanced capabilities to achieve full maturity, from DevOps processes with GitHub Actions or Azure DevOps pipelines to multiple environments or automated test execution with Test Engine.

How test planning benefits your users

In the last part of this book, we will dive deeper into telemetry and Power Apps and go through all the available services, both basic and advanced. However, in our citizen developer scenario with the calculator app, we wanted to show the importance of monitoring with some examples.

Monitoring helps you uncover bugs and improve your tests for the app's functionality. This extends to understanding your users' needs, gaining insights from their feedback, and leveraging monitoring tools to optimize your app's performance. To put it simply, the more accurately you can mirror real-world use in your testing, the better your app will function for your users when it matters.

For instance, we will delve into tools such as **Power Apps Monitor** and **App Analytics**. Power Apps Monitor allows you to track and inspect the behavior of your apps, helping you understand how users interact with your apps, and where potential bottlenecks may occur. This information is invaluable when defining test cases and scenarios as it allows you to anticipate problems and address them proactively.

Likewise, App Analytics provides detailed usage statistics, helping you understand the features users interact with most, peak usage times, and more. Such insights can help tailor your testing approach to ensure that high-demand features function flawlessly and that the app can handle traffic spikes smoothly.

Using Power Apps Monitor for real-time debugging

To effectively test and troubleshoot your calculator app, you should use Power Apps Monitor, a real-time debugging tool that provides an in-depth look at your app's behavior as you run it. Power Apps Monitor enables you to gain detailed insights about the operations that are taking place in your app, which can help you identify and address both performance issues and accidental errors.

To initiate a monitoring session, open Power Apps Monitor before you launch your app through the **Advanced Tools** option and choose **Open Monitor**. Now, when you run your calculator app, Power Apps Monitor will record the operations in the background, giving you a comprehensive under-the-hood view.

Figure 4.9 displays telemetry entries for each user interaction. Selecting a single entry, such as the one highlighted, from the **UserAction** category, allows us to understand how it was executed, the data sources involved, and more:

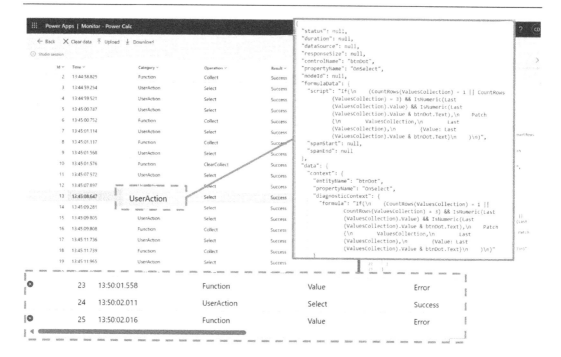

Figure 4.10 – Monitor showing telemetry from Power Calc

Moving into lines 23 and 25, an error signal shows up, with an internal message:

```
"errorMessage": "The value '1213102.8.' cannot be converted to
a number."
```

The user tries to add a double decimal sign, and it is managed internally with an error. Looking into it, and to avoid errors, we could review the associate code and disable the button until an operator is clicked.

One of the main advantages of Power Apps Monitor is its collaborative nature. You can invite other users to view the application monitor data, making it a great tool for team-based troubleshooting. If you encounter an issue that you are not sure how to fix, you can share the data with a more experienced developer or team member and ask for guidance.

In practice, this could be as simple as sharing the session with your fictional colleague when your calculator app takes too long to perform complex calculations. With the granular data provided by the monitor, they can pinpoint whether the issue stems from the formula logic, data source delay, or an inefficient control property.

When you invite someone by using the **Invite** option from the top menu, they should have access to the same environment and will only see telemetry entries from the moment you started sharing the monitor, and for a maximum of 60 minutes. In our case, both users can download the telemetry data in JSON format. This is interesting for offline analysis such as importing the data into Excel.

Downloading and uploading data with Power Apps Monitor for efficient issue tracking

Power Apps Monitor not only provides real-time debugging capabilities but also allows you to download and upload session data. This feature is incredibly useful in keeping track of issues and validating fixes over time.

The exported file has two main elements:

```
{
    "Version": 2,
    "SessionId": "546754b0-294e-11ee-9545-897f12d69594",
    "Messages": [],
    "Config":[]
}
```

The `Messages` array stores each telemetry log from user interaction, while `Config` contains editor and engine configuration data. In Excel, follow these steps:

1. Select **Data** | **Get Data** | **From File** | **JSON** and select the downloaded Monitor file. You could use any application for this – in our case, we'll use the calculator application from this book's GitHub repository and import it into our environment.

2. This will open **Power Query Editor**, where you need to expand the columns to get the data. Just click on the dropdown at the column name.

3. Finally, remove the unwanted columns; check out *Figure 4.10* to see the process. You can check the Excel file in this chapter's GitHub repository, which includes sample data from Monitor telemetry logs from the calculator app and sample analytics:

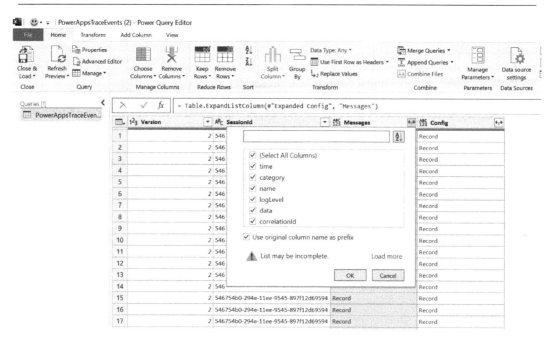

Figure 4.11 – Monitor telemetry imported into Excel

You can analyze the data through some simple pivot tables. As an example, check *Figure 4.11* to quickly see the distribution of app activities and errors. We can see that users spend more time clicking numbers, but we do not know how much in terms of operators. In that case, we could add a custom trace to differentiate between them:

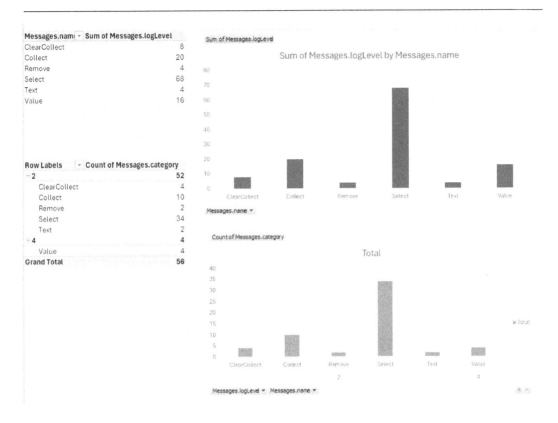

Figure 4.12 – Monitor telemetry reports

You can then use this data for offline analysis, but you can attach it to an issue you need to fix.

For example, let's say the calculator app has a recurring issue where certain calculations result in errors. You can run a monitor session, replicate the error, and then export that session. The exported data can then be attached to the issue in your issue tracking system. This gives anyone working on the issue a clear snapshot of what was happening in the app when the error occurred.

Moreover, the upload feature allows you to load previously saved sessions back into Power Apps Monitor. This can be particularly useful when you want to verify if a bug has been fixed. By uploading the saved session that initially caught the error, you can compare it side by side with a new session after applying the fix. This way, Power Apps Monitor helps streamline your test planning and validation process, making troubleshooting your calculator app more efficient and effective.

Power Apps Analytics

From a citizen developer's perspective, Excel gives you the flexibility to drill down into the monitor logs, but if you want to get an overview of the overall app analytics, and you are not the administrator (Environment, Microsoft 365, and Power Platform), you can still get information at the app level.

To check it, you need to navigate to https://make.powerapps.com, select **Apps**, and look for your app. Once you've done this, an **Analytics (Preview)** option will appear with three different reposts:

1. **Usage**: This dashboard is the default when you open the analytics page and gives you the distribution of app launches and active users. *Figure 4.12* shows the menu you can use to export data for each graphic in a summarized or raw data format:

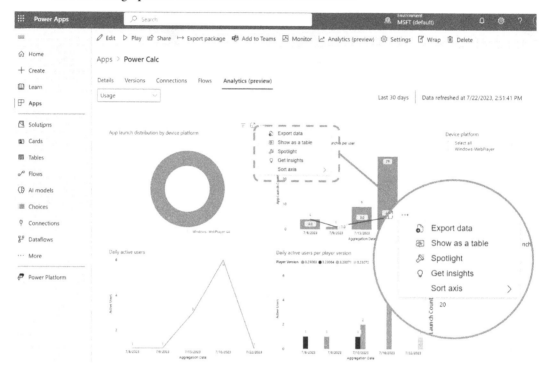

Figure 4.13 – The Power Apps Analytics Usage page

2. **Performance**: This dashboard summarizes key performance metrics for Power Apps: the time taken from app launch until control is handed over to the app's first screen (including and excluding connection setup time), the duration users spend in the app per session, and the user distribution across different session lengths. *Figure 4.13* shows an example of the performance of the app and gives us information about how the users use it. The users spend little time in the app by far (less than 1 minute) versus spending more time 5 minutes. We could try to validate the assumption that exporting previous calculations could help our users to use it more effectively by reusing it in other scenarios such as Excel import:

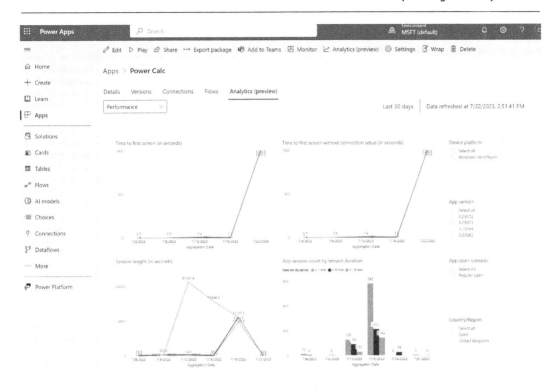

Figure 4.14 – The Power Apps Analytics Performance page. Note that this
screenshot only intends to show the structure of the page

3. **Location**: The final dashboard helps you understand the global distribution of your app. It
    makes sense if you have a distributed workforce and you need to understand the potential
    impact of new versions, or potential localization issues. For example, *Figure 4.14* shows a map
    where all activity was undertaken in Europe, Spain:

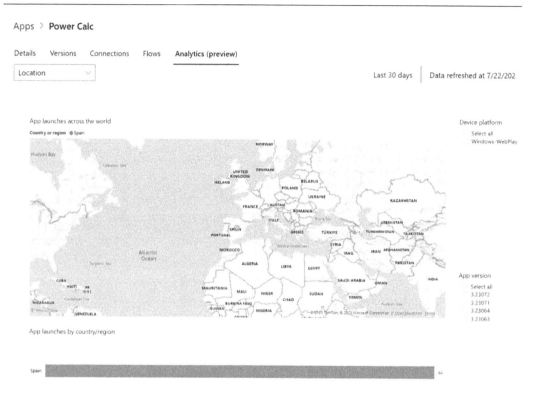

Figure 4.15 – The Power Apps Analytics Location page

The reports are Power BI embedded reports, so they bring functionality out of the box, such as export data you can explore on your own.

Congratulations on completing this chapter on testing strategies for Power Apps development! You have come a long way; let's summarize our learnings.

Summary

In this chapter, we dissected the process of applying testing concepts to planning a Power Apps development project. We delved into how the clear definition of an app aids in outlining testing activities, the influence of an organization's maturity on planning, the transition from individual developers to fusion teams, and the advantages of test planning for users. By mastering these topics, you've gained a holistic understanding of how testing strategies guide app development, and how collaboration within a fusion team can amplify the project's effectiveness.

As we advance, we will start *Part 2* of this book, where we will deepen our exploration of Power Apps tools such as Test Studio and Test Engine. We will discuss how these capabilities can aid in optimizing the app development and testing phase, thus refining your project planning strategies even further.

Part 2: Tools for Power Apps Automated Testing

In *Part 2*, we will explore the key tools used in Power Apps testing. We will start by covering Test Studio, with both an overview and a practical example. We will then switch focus to Test Engine, discussing its evolution, comparisons, use cases, and new improvements. The part wraps up with a chapter on testing canvas apps, using Power Automate Desktop as an alternative for certain scenarios.

This part has the following chapters:

- *Chapter 5, Introducing Test Studio – Canvas Apps Testing with a Sample Case Study*
- *Chapter 6, Overview of Test Engine, Evolution, and Comparison*
- *Chapter 7, Working with Test Engine*
- *Chapter 8, Testing Power Apps with Power Automate Desktop*

5

Introducing Test Studio – Canvas Apps Testing with a Sample Case Study

Testing and verifying are integral components of the software creation process. As we have seen in previous chapters, it allows us to identify problems or defects throughout the entire software life cycle, ensuring the quality of the application. This is also applicable to applications created with Power Apps. Depending on the size and number of users of the application, it may be sufficient to manually test it, but in most instances, we need to consider automated testing strategies instead of manual testing.

In this scenario, we will begin to consider implementing automated testing cycles and regression tests. This automation will bring the next set of benefits, among others: greater accuracy, time and cost savings, and improved quality. It will also enable the implementation of continuous testing processes as part of DevOps, bringing more flexibility. Most importantly, it will allow for test reuse and scalability by providing instant feedback.

Using Test Studio, you can create a series of tests that simulate user interactions with the app, such as filling out a form, clicking a button, or navigating to a specific screen. You have the option to compose these tests by utilizing Power Apps expressions (Power Fx) or employ the recorder to generate tests based on your interactions with the app.

Once you have created these tests, you can replay them to verify that the app is functioning correctly and that users can interact with it as intended. For example, you might run a test that simulates a user entering information about a new book and saving it to the app. If the test runs successfully, you can be confident that the app is functioning correctly and that users will be able to create new book entries.

In this chapter, we will learn how to familiarize ourselves with the Test Studio interface and how to write tests for Test Studio using Power Fx expressions or the recorder. We will use an example Canvas App that you can find in the GitHub repository provided or create while following this chapter's content.

This chapter consists of the following topics:

- **Test Studio overview**: This section introduces the terminology associated with Test Studio, discusses the best practices for using it, and outlines its limitations.

- **Working with Test Studio, test functions, and the recorder**: we will guide you through accessing the Test Studio interface and familiarize you with its primary components, including the recorder.

- **A first example of Test Studio with a Canvas App sample**: this section provides a hands-on demonstration using Test Studio with a sample Canvas app specifically designed for this book.

- **Processing and saving test results**: this part deals with managing test results. It includes instructions on how to save the results for future analysis.

Technical requirements

To follow the examples in this book, you can download the corresponding files from the designated GitHub repositories. We will utilize two repositories:

- The example and code repository provided by Microsoft for the Test Engine component, which will be discussed in the upcoming chapter: `https://github.com/microsoft/ PowerApps-TestEngine`.

- The GitHub repository, which was specifically created for this book, can be accessed at `https://github.com/PacktPublishing/Automate-Testing-for-Power- Apps/tree/main/chapter-05`. We will reference the elements contained within this repository later in this chapter.

To clone the repositories, you will need to have Git installed on your computer. If you don't have it installed already, you can download it from the official Git website (`https://git-scm.com/ downloads`).

While not strictly necessary for this chapter, having a code editor such as Visual Studio Code can significantly help you with editing certain components. You can download Visual Studio Code from `https://code.visualstudio.com/`.

Access to a functioning Power Platform environment is necessary to install the provided examples or create new ones. The Power Platform developer environment appears to be the most suitable solution for this purpose. Later in this chapter, we will explain how to obtain access to such an environment.

Test Studio terminology

In the world of test automation, several terms are commonly used to describe the elements that are part of it.

Among them, all of which are related to Test Studio, we have the following:

- **Test cases**: When developing an application, we segment it into various use cases, which essentially represent the different functionalities we aim to incorporate within the application. These use cases form the basis of our test cases.

  For example, if we are working with an expense tracking app, we need to ensure that only allowed expenses can be submitted for a given user profile. In this case, the test case would have to cover the necessary interactions to demonstrate that this requirement is always met. Test cases are composed of a series of steps, which we will call *test steps*. These test steps are written using Power Fx, the coding language that's used across Power Platform (`https://learn.microsoft.com/en-us/power-platform/power-fx/overview`).

- **Test suites**: Test suites serve as a means to arrange and categorize the test cases. When the application is of considerable size, the number of test cases can also be considerable, so it is convenient to organize the test cases by the features or functions we are dealing with. For example, you can have a test suite to validate the expense reports we mentioned earlier and organize another test suite to validate purchase orders within an organization.

  In these test suites, the test cases are executed one after another, with the state from earlier test cases being preserved during the execution of subsequent test cases. This is very useful for simulating conventional navigation through the application. For example, if we have a test case that is waiting for a login operation on the home screen, the test suite should start the process on the login home screen; then, the next test cases can use the login performed in that operation. The application is not reloaded at the start of each test case of a suite – that is, it maintains the state of the previous test case.

- **Test assertions**: A test assertion is a statement that describes the expected outcome of a test case. They are used to verify that the expected behavior of the application is being met.

  For example, let's say we have a test case that checks whether a user can add items to their shopping cart on a Canvas App. The test assertion, in this case, might be "Item X was added successfully to the shopping cart." If the actual outcome of the test case matches the expected outcome described in the test assertion, then the test is considered to be successful. However, if the actual outcome does not match the expected outcome, the test is considered to fail, and the error is reported. This allows us to quickly identify and fix any issues with the shopping cart functionality of the app. It's important to acknowledge that each test case can contain multiple assertions.

Having outlined the key terminology employed in the world of Power Apps test automation, let's delve into the best practices that should be considered in this context.

Best practices

When testing Canvas Apps using Test Studio, it is advisable to consider the best practices that will allow us to get the most benefit from our automated testing. Let's look at some key ones:

1. First of all, it is *important to determine which test cases we should automate*. Although it is generally recommended to automate most tests, there are cases where automation may not be the most convenient mechanism, and in these cases, a manual test may be more practical. Here are some examples:

   - Tests that are only performed infrequently, such as those for very specific edge cases.

   - Tests that require a high level of flexibility and adaptability, such as those that involve complex user interactions or visual elements.

   - Tests that require a high level of human judgment, such as those that involve verifying the quality or accuracy of content.

   - Tests that are very time-consuming to automate, such as those that involve a large number of steps or complex integrations setup. However, even in this scenario, it may be beneficial to invest in the test automation process if the tasks are performed regularly. Ultimately, it's a matter of balancing the required investment with the potential return on investment.

2. *Keep test cases small* – it is not recommended to write a large test case that covers all the functionality of the application. Instead, it is advisable to break it down into several test cases, with each one covering a distinct feature of the application. This will make it easier to validate changes and facilitate the management and maintenance of the test cases. Additionally, *smaller test cases are typically easier to understand and debug*, which can save time and effort in the long run.

   Let's say you are testing a calculator app, and you have a test case to check that the addition functionality works correctly. However, if there is an incorrect assertion in that test case, it could cause another functionality (such as subtraction) to remain untested. To prevent this from happening, you can group your test cases into a test suite. By doing so, even if the "addition" test case fails, the other test cases in the suite (such as subtraction, multiplication, and division) can still be run and validated. This ensures that the automated testing of the calculator app is thorough and accurate and that all of its functionalities have been properly tested.

3. While it is possible for a test action to include multiple expressions, having multiple actions within a single step can negatively impact the ability to identify and resolve test failure. A more effective approach is to *split a test step with numerous actions into multiple test steps* and assign a single action to each, which will make it easier to identify issues quickly. This will help ensure that the automated testing is efficient and effective and that any problems can be quickly identified and resolved.

4. *Each test case must have at least one expected outcome.* Test assertions will allow us to validate the expected results against the actual results. When anticipating various results, you can include multiple assertions within a single test case. This will help ensure that the test case covers all the expected behavior of the application and that any deviations from the expected behavior are detected and reported accurately. By using test assertions, we can make our automated testing more comprehensive and reliable. The test case outcome will be based on the expected combination of one or multiple test assertion results.

5. *Use test suites to organize similar test cases for better maintenance and understanding.* This will help ensure that your automated testing is efficient and effective and that you can quickly identify and resolve any issues that may arise.

As we begin to work with Power Apps Test Studio, it's important to understand that it currently has some limitations in terms of functionality. We will briefly discuss these in the following section.

Limitations

At the time of writing this book, Power Apps Test Studio lacks support for certain functionalities. These include, but are not limited to, the following:

- Components (both canvas and PCF)
- Nested galleries
- Media controls
- Support for controls not listed in the `Select` and `SetProperty` functions

In the subsequent chapters, we will delve into how we can circumvent some of these limitations using **Test Engine**. For a comprehensive and updated list of these limitations, please refer to `https://learn.microsoft.com/en-us/power-apps/maker/canvas-apps/test-studio`.

> **Note**
> It is worth noting that work is ongoing to provide full control coverage in Power Apps Test Studio.

Preparing the environment for Test Studio

Are you ready to get started with the Test Studio environment? Great! Before we begin, we need to set up a Power Platform working environment.

Fortunately, we have two options to choose from: the Power Apps Developer Plan or the Power Apps Trial Plan.

> **Note**
>
> The Power Apps Developer Plan is the most convenient option as it's a free development environment that gives you access to all the full-featured Power Apps and Power Automate development tools, as well as the Dataverse data platform. All you need is a work or school email address to sign up for the Developer Plan.

Signing up for the Power Apps Developer Plan

The Power Apps Developer Plan offers a free development environment for creating and testing Power Apps and other Power Platform components. To sign up, follow these steps:

1. Go to the Power Apps Developer Plan website (`https://aka.ms/PowerAppsDevPlan`) and log in using your work or school credentials.

2. Click the **Get Started Free** button on the web page.

3. Once you've signed in, and if you have the needed permissions given by your organization, you will be directed to the developer environment that has been created for your user.

To obtain additional support for the Power Apps Developer Plan application process, please visit the Power Apps Developer Plan page at `https://learn.microsoft.com/en-us/power-apps/maker/developer-plan`.

If you find yourself unable to utilize the Power Apps Developer Plan – perhaps because your organization has disabled this feature – you might want to consider contacting your organization's supportteam. Alternatively, you can still tap into the resources offered by the Power Apps Trial Plan. This 30-day free trial provides temporary access to all the features of Power Apps.

Power Platform trial environment

To sign up for a 30-day Power Apps Trial Plan and explore all the capabilities of the platform, you can follow these steps:

1. Go to the Power Apps website (`https://powerapps.microsoft.com/`) and log in with your work or school credentials.

2. Click on the **Try free** button on the home page, or if you are using a phone, click on the menu in the top-right corner and select **Try free**.

3. In the pop-up window that appears, enter your work or school email address and click **Submit**.

4. If a dialogue box appears indicating that Power Apps recognizes your organizational credentials, follow the prompts to complete the sign-up process. If you do not see this dialogue box, check your email for a verification link and follow the prompts to verify your email address.

Once you have completed the sign-up process, you will have temporary access to all the features of Power Apps for a 30-day trial period. If you do not extend your trial or purchase a plan after the trial period ends, you will be prompted to request an extension or purchase a plan. You can extend the trial an additional two times, with each extension lasting 30 days. The maximum trial period is 90 days in total. If you have another type of license, you will still be able to use Power Apps features that are covered by your other license. Any data in Dataverse will be preserved and any app or flow that uses Dataverse will continue to function, so long as your license supports it. However, if you attempt to use premium Power Apps features that are not covered by your existing license, you will be prompted to purchase a plan. You can find more information about the different plans on the pricing page.

Now that we have our environment, we will describe the process for downloading or cloning the repositories that we will be using, and subsequently describe how to import the first example solution that we will use for this chapter.

Setting up the Git repository

Follow these steps to clone the repository:

1. Open a terminal window on your computer.
2. Navigate to the directory where you want to clone the repository. You can use the cd command to change directories.
3. Run the following command to clone the repository:

    ```
    git clone https://github.com/microsoft/PowerApps-
    TestEngine.git
    ```

    For the GitHub repository for this book, run the following command:

    ```
    git clone https://github.com/PacktPublishing/Automate-Testing-
    for-Power-Apps
    ```

4. The repository will be cloned to a new directory called PowerApps-TestEngine for the test engine one or Automate-Testing-for-Power-Apps for the dedicated repository for this book.

> **Note**
> Alternatively, you can download the repository as a ZIP file from the GitHub website. To do this, in the repository pages, click the **Code** button, and then click the **Download ZIP** button.

Now that we have downloaded both repositories, in the next section, we will cover how to import the solution for this book's GitHub repository into our test environment.

Importing the solution

Within the GitHub repository for this book, locate the `Chapter5_1_0_0_1.zip` file. This ZIP file contains the **Gallery Application**, which we'll be using as our example. To import the solution, follow these steps:

1. Go to `make.powerapps.com` and navigate to the **Solutions** page in the Power Platform developer environment. You can find this page by clicking on the **Solutions** tab in the left menu.

2. Click the **Import** button at the top of the page.

3. In the **Import a solution** window that appears, click the **Choose file** button and select the ZIP file containing the solution (found in this book's GitHub repository).

4. Click the **Import** button to begin the import process.

5. The import process may take some time, depending on your internet connection speed. Once the import is complete, the solution will appear in the list of solutions on the **Solutions** page.

In the next section, we will outline the process of importing a small dataset for our sample application, enabling it to execute test operations.

Creating data demo

To work with the application, we will use the **Dataverse "Contact" table**. This table is likely empty in the development environment we have created, so we will import some sample data into it.

> **Note**
> An Excel file with 50 sample contacts is provided in this book's GitHub repository, but you can use any other file that is more familiar to you.

To import the file, perform the following steps:

1. Open a web browser and go to `make.powerapps.com`.

2. Sign in with your Power Apps account and select the appropriate environment in the toolbar.

3. Click on the **Dataverse** menu in the left navigation bar.

4. In the **Dataverse** menu, click on the **Tables** option.

5. This will bring you to the **Tables** page, where you will see a list of all the tables in your **Dataverse**. Locate the **Contacts** table in the list and click on it to open it.

6. The **Contacts** table should now be displayed, showing any existing data that is already stored in it. If it is empty, proceed to the next steps.

7. Click on the **Import** button in the top menu and select **Import data from Excel**.

8. Click on **Upload** and choose the democontacts.xls file from your computer.

9. Click on the **Map columns** button to proceed to the mapping step.

10. Assign the unmapped columns: map **Country** to **Address1: Country/Region**, **Phone** to **Address1: Phone**, and **Email** to **Email**. After mapping, click **Save Changes**:

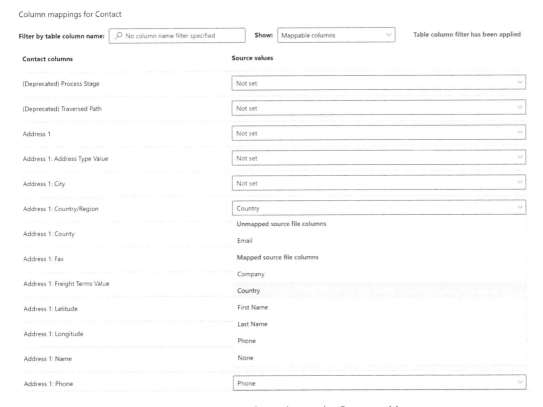

Figure 5.1 – Mapping demo data to the Contact table

11. The mapping status message should say **Mapping was successful**.

12. Click on the **Import** button to start the import process.

13. Wait for the import process to complete. A notification will be displayed, indicating the number of records that were imported successfully.

14. Click on the **Close** button to finish the import process.

The imported data should now be available in the **Contact** table, as shown in the following figure:

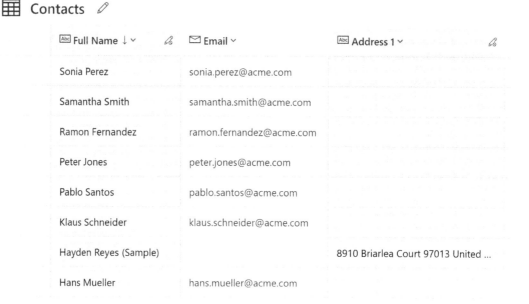

Figure 5.2 – Contact data imported

Thus, with our environment and solutions prepared, we are ready to begin working with them. Let's proceed to the next section, where we will start testing our application.

Our first example with Test Studio

Let's dive into Test Studio and explore the sample application hands-on. In the following section, we will describe the process of creating the application to be used. We will utilize the gallery template, which is derived from the gallery control (`https://learn.microsoft.com/en-us/power-apps/maker/canvas-apps/controls/control-gallery`), to showcase and navigate the **Contact** table.

If you prefer to bypass these steps, the completed app is available within the solution provided in this book's GitHub repository, as mentioned previously.

Creating the app

Now that we have the sample data, we can create the canvas application that we need to test in Test Studio. To do this, we will use a simple gallery app, so let's start creating it.

To create a canvas app with the gallery template, follow these steps:

1. Select the **Solutions** option from the left navigation bar, followed by **New Solution** from the top bar. Provide a display name for your solution (an internal name will be automatically generated), choose the default publisher to enable, and click the **Create** button.

2. After creating the solution and navigating inside it, you can create your app. To do this, click on the **New** button in the top bar, select **App**, and then choose **Canvas App**.

3. Assign a name and choose the tablet format for more space. Click **Create**.

4. When the app editor appears, click on the **Create a gallery** link that is shown in the popup.

5. A new popup will appear with **Select a Data Source**. Here, we can choose our **Contacts** table.

6. Adjust the gallery control so that it occupies half of the screen and drag a label control to the other half. Format it and assign `Gallery1.Selected.'Full Name'` to the **Text** value.

7. Repeat these steps with **Email**, **Address1: Phone**, and **Address1: Country/Region**:

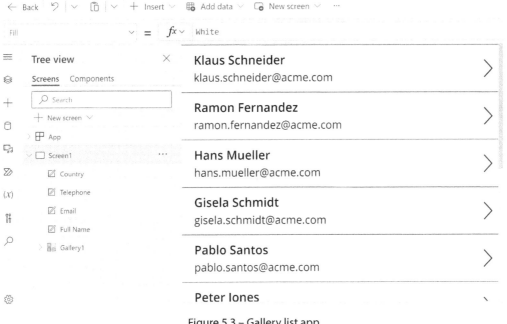

Figure 5.3 – Gallery list app

8. You can now test your app by clicking on the **Play** button in the top menu or *F5* on your keyboard. This will open a preview of the app in your web browser.

9. Once you are satisfied with your app, you can save it by clicking on the **Save** button in the top menu and/or publish it by clicking on the **Publish** button in the top menu:

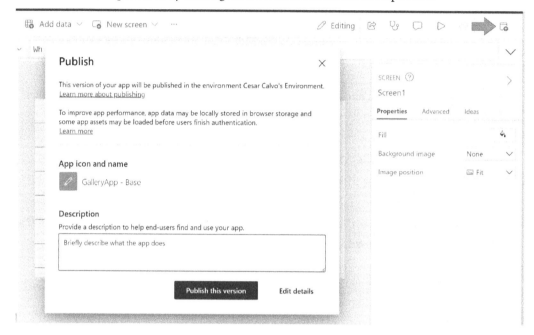

Figure 5.4 – Publishing the application

With our application set up, let's proceed to Test Studio to initiate the testing process.

Accessing Test Studio

To begin testing with Test Studio, we will open the application that we have created or imported from the solution in this book's GitHub repository.

Once the Power App has been saved, which is a requirement before we can start writing tests, we must select **Advanced tools** in the left navigation menu, as shown in the following screenshot:

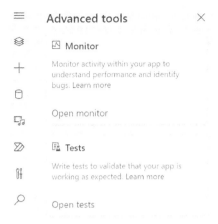

Figure 5.5 – Advanced tools – Tests

Here, we will find a tile labeled **Tests** where we can click on the **Open tests** link. This is where we will find the Test Studio interface and we can begin exploring:

Figure 5.6 – Sample app included – Tests

Note

Once we've developed and published our tests, the tests will be saved with the app package. As a result, whenever we export and import the app into another environment, all the test definitions, suites, and cases we have created will also be included. For example, this is what you will find if you are using the app included in the sample files.

Before delving into the creation process, let's briefly outline the components that make up the Test Studio interface.

The Test Studio interface

The components available in the main interface of Test Studio will differ based on the option you choose. Here are the key elements and their locations:

- When selecting a Test Case or Test Suite, you have the **Download Suite** option in the top bar. This feature will become quite useful when we discuss the new Test Engine option in the upcoming chapter.

- When selecting a Test Case step, the top menu bar of Test Studio offers quick access to create new steps of various types, such as **Assert**, **SetProperty**, **Select**, and **Trace**. We'll delve into these types in subsequent sections:

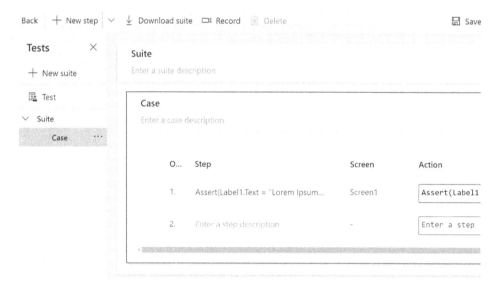

Figure 5.7 – The Test Studio interface

- When selecting the Test Case, you'll also find the **Record** button. This allows you to record the steps for inclusion in your test suite through actions performed in the application.

- When selecting a step within a Test Case, you'll find menu actions for **Insert Step Above** and **Insert Step Below**. Additionally, you'll see an option to **Delete** the step.

- In all cases, to the right of the top menu bar, you'll find the usual options for **Save**, **Publish**, and **Play** (execute these operations to observe the behavior of the created tests). There's also an option to copy the reproduction link for use in different processes, such as DevOps processes, which we will cover in a later chapter.

The best way to familiarize yourself with the interface is via a practical example. This is something we'll do in the next section.

As I mentioned earlier, we will divide the tests into test suites, which are the sets of tests we want to group as an end-to-end operation, the test cases, which are the sets of steps we will execute in our tests, and finally, the steps that will be part of the steps covered in each of the test cases.

As we can see, *we can start creating a new suite from the left column*, or use the one already created by default, simply by renaming it through the three dots that appear to the right of the word **Suite**. The same goes for the default created test case, and in each of the test cases, we can start creating the steps for these cases.

As indicated, we can create the steps manually through Power Fx expressions, or we can use the recording tool to record the steps that we need in our tests.

> **Note**
>
> The ideal approach may be a combination of both – that is, using the recorder once we have saved our application for movement operations through the interface and using the Power Fx syntax for assertions, traces, and variable assignments.

Creating a test suite

To create a new test suite, you can follow these steps:

1. Select **New suite** from the options available. To update the name and description of the test suite, choose the appropriate fields on the main grid and modify them accordingly.

2. Enter the desired name and description for the test suite. You can do this in the **Suite** and **Enter a suite description** fields.

3. Save your changes by clicking the **Save** button:

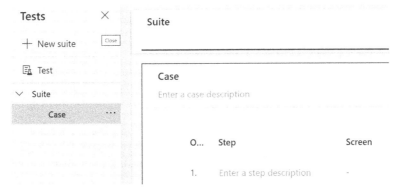

Figure 5.8 – Test suite creation

> **Note**
>
> You can also rename the existing test suite that was created by default through the three dots that appear to the right of the word **suite**.

Creating a test case

Now, let's start creating a test case. We can decide to create one or several test cases within a test suite, depending on how we wish to structure or group our tests. It is important to keep the previously stated best practices in mind, such as keeping the test cases simple and targeted to test a particular feature or functionality within the application.

To start creating a test case, *we will first select the suite to which we want to assign it*. For now, we will select **New case** in the top menu to create a new case, though we can rename the existing one. We will update the name and description according to the functionality we want to test with this test case:

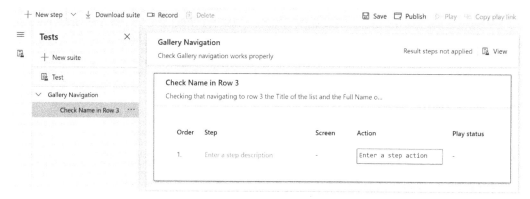

Figure 5.9 – Test case creation

To begin creating the steps for our test case, we must utilize the recording functionality built into Test Studio.

Recording a test case

We will start with the process of recording a test case.

> **Note**
>
> Recording mode can only be performed on published applications. Therefore, before recording a test case, ensure that any recent changes to the application have been published. Otherwise, recording without publishing these changes could result in testing an outdated version of the application, leading to inaccurate test results.

Let's select **Record** from the top menu. With this action, the previously published application will open in recording mode in a new browser tab.

> **Note**
>
> It is important to remember that this recording will overwrite any of the steps of the test case that are present at that time.

The final intention of this test is to verify that when we navigate to one of the records in the list, the data that appears on the detail screen corresponds to the selected record, meaning that navigation is working correctly. Follow these steps to interact with the application:

1. Select row 3 in the gallery list – that is, the name of the contact.

2. In addition, select the full name that appears on the application's detail screen, which should correspond to the same name that appears in the selected row.

3. Now, select the subtitle of row 5 (that is, the email). On the right-hand side of the screen, select the corresponding field for the email in the detail area.

If we click the **Done** button, we will save the recorded steps by assigning them to the test case. We also have the option to click the **Cancel** button, in which case the recorded steps will not be saved. Once we click **Done,** we will be able to see the recorded steps and the expressions that have been automatically generated by Test Studio:

Figure 5.10 – Recording mode

As can be seen in the preceding screenshot, in this case, the **Title1** field of the list and the **Label1** field of the detail screen correspond to the name values, and the **Subtitle1** field of the list and the **Label2** field of the detail screen correspond to the email values.

On the left-hand side of the screen, you'll find the steps that you've recorded. Please verify these steps as we'll be referencing them in the forthcoming sections.

We can edit the description of the test if necessary and we can also update the actions of the steps by selecting the formula and editing it. In the next section, we will see how to create new steps and assertions based on the Power Fx syntax.

Adding test steps with Power Fx

In this section, we will learn how to add additional steps manually to our Test Studio test, both to trace the actions being executed in the tests and to check or assert the values that we are expecting as a result of the test. Before that, we will delve deeper into the operations available in Power Fx that are most suitable for test execution, in addition to the other functions available in Power Apps:

- The `Select` function in Power Apps mimics a user's select action on a control, such as a gallery. For instance, to propagate a select action to a parent control like a gallery, we can use the `Select` function. The practical application of this function will be illustrated in the upcoming sections and chapters. For more information on how to use the `Select` function, refer to the official reference documentation available at `https://learn.microsoft.com/en-us/power-platform/power-fx/reference/function-select`.

- The `SetProperty` function in Power Apps Test Studio mimics user interactions with input controls by setting values on their properties. This function is exclusively available when writing tests in Power Apps Test Studio. For more details on using the `SetProperty` function in your tests, consult the official reference documentation at `https://learn.microsoft.com/en-us/power-platform/power-fx/reference/function-setproperty`.

- In Power Apps Test Studio, the `Assert` function is employed to verify that the actual result of a test or test step matches the expected result. If the expression evaluates to false, the test case will fail. Assertions can be used to validate control properties such as label values, list box selections, and other controls within the app. For further details on how to use the `Assert` function in your tests, please refer to the official reference documentation available at `https://learn.microsoft.com/en-us/power-platform/power-fx/reference/function-assert`.

- The `Trace` function is used to add supplementary information to your test results from the `OnTestCaseComplete` event, as we will see later. Suppose you have created a test that includes a sequence of steps to validate a registration form. You could use the `Trace` function to add diagnostic information to the test results at specific points during the test's execution. For instance, you might add a trace message before the test verifies the user's email address to confirm that the email validation is occurring correctly. For more information on using the `Trace` function in your tests, please consult the official reference documentation at `https://learn.microsoft.com/en-us/power-platform/power-fx/reference/function-trace`.

Figure 5.11 – Insert steps

In the test case that we have already created, we will proceed to delete the steps related to **Label2** and **Subtitle1** since we will add these steps in a new test case.

We will create an additional test case (by selecting our Test Suite and clicking on **New case**), in which we will check that the subtitle we have in the selected element of the gallery corresponds to the email we have on the details screen. Therefore, we will name this test case `Check Email in row 5` and we will add the steps referring to selecting **Label2** and **Subtitle1** that belong to the email field check in the selected record of the gallery list (we have chosen row 5 for the test). We can do this through a new recording in this test case or simply by adding the corresponding `Select` functions. You can modify the name of a test step by hovering over its name and clicking on the pencil icon.

We can see the additional test case that was created in the following screenshot:

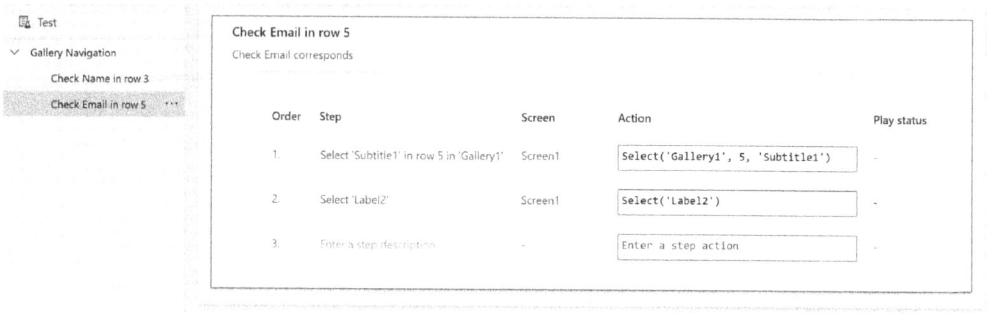

Figure 5.12 – Additional test case

Now, coming back to our first test case (the one related to row 3 named **Check Name in row 3**), we are going to add a new step (for instance, below the initial step). In this step, we will cover the assignment of the value of the **Title1** field to a variable that we will call `Title`.

To add a new test step, choose the step that was placed after the desired location for the new step; after that, pick **Insert a step above** from the primary menu or the active row. This will generate a blank step for you to personalize.

The syntax for variable assignment in this case would be as follows:

```
Set(Title, Gallery1.Selected.Title1.Text)
```

We can name this step `Assignment of value to variable title`.

Now, let's add a `Trace` function step to show the information stored in this variable. Again, this is not necessary for executing the test in this case and is simply intended to demonstrate this functionality.

Let's name this step `Trace of the variable title`. The syntax for this step will be as follows:

```
Trace("Variable Title is: " & Title);
```

Select the **Label1** label (or the alternate name that has been chosen for this label) on the detail screen:

```
Select(Label1)
```

Finally, let's add an `Assert` function where we will check that the result after selecting the record of row 3 is that both the variable title and the value of the **Label1** label match. The syntax for this function is as follows:

```
Assert(Label1.Text = Title, Title & " does not correspond with " &
Label1.Text)
```

The name we can give to this step could be something like **Final Assert**. And a reminder, in the syntax for the `Assert` function, `Assert(expression, message)` - `message` is an optional description of the assertion failure. The following screenshot shows where we can check the final result for our **Check Name in row 3** step:

Figure 5.13 – Test case with an assert

Now, let's go to our other step, **Check Email in row five**. We're going to modify the elements that were placed there as well.

In this case, we previously selected the subtitle of element number 5 in the gallery list (which corresponds to the email) and the **Label2** label on the detail screen, which also corresponds to the email value in the elements we show on the detail screen.

We will add a step that corresponds to the *assignment of a variable* to the **Subtitle1** element that we have selected for this. We can use a syntax similar to this:

```
Set(Email, Gallery1.Selected.Subtitle1.Text)
```

After this variable assignment, we will add an additional step for the *trace of this value* that we have assigned. It is important to remember that these trace values will not be visible on the screen during the test playback, but we can export them to the results table, as we will see in the next section. The syntax for this trace would be something like this:

```
Trace("Variable Email is: " & Email)
```

Finally, after selecting **Label2**, we will add the *final assertion* to check that the values match, just as we did with the previous test case:

```
Assert(Label2.Text = Email, Email & " does not correspond with " &
Label2.Text)
```

The result of our modifications in this step should match what's shown in the following screenshot:

Figure 5.14 – Test case – assert email value

With that, our tests are ready. Let's start the playback!

Playing back tests

To validate the app's functionality, let's play back the recorded tests. Remember to publish your app before you play back the latest changes. Publishing the application ensures that the latest code and resources are included.

After clicking the **Play** button, we will be able to watch the tests being run and view the results. The results in *Figure 5.15* were obtained by executing the entire test Suite. However, it's also possible to run test steps individually:

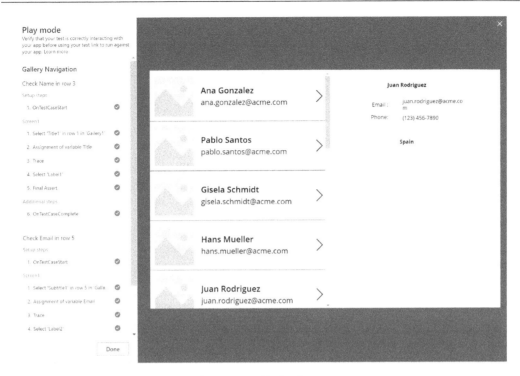

Figure 5.15 – Playback tests

Once the tests are completed, click on **Done** to finalize the execution. In the next section, we will explore additional properties that can be customized.

Setting up tests

The `OnTestCaseStart`, `OnTestCaseComplete`, and `OnTestSuiteComplete` properties can be customized to initialize and transmit the outcomes of your tests to alternative data destinations. For instance, the `OnTestCaseStart` attribute of a Test Suite can be employed to prepare the test if you execute tasks shared among all test cases within the suite. You can do this as follows:

1. Navigate to any screen of the application and initialize it in a certain state.

2. Fetch test data for the current test.

3. Initialize variables that will be used in the testing process.

First, we will configure the `OnTestCaseStart` property to set a starting value for a variable and ensure that the tests always start on the screen that we want to test. This last case does not apply to the example we are using, but it will be very convenient for other types of applications. To configure this property, we can click the **Test** button in the left column or the **View** button on the main screen. When we click on one of these two buttons, we will find the `OnTestCaseStart`,

`OnTestCaseComplete`, and `OnTestSuiteComplete` properties, of which we will only deal with the first one, `OnTestCaseStart`, for now. We will see the other two in the next section.

First, we will configure the application to always start on the screen where `GalleryList` is located. To do this, we will use the `Navigate` function:

```
//Start every case on the screen with the gallery list
Navigate(Screen1);
```

Now, we will declare a variable that contains the total number of records that `GalleryList` presents at the beginning of the tests. The final idea is that during the test process, this number of records can be checked to ensure that it is not altered:

```
// Count the total number of items that are in the GalleryList
Set(totalContacts, CountRows(Gallery1.AllItems))
```

The result of our `OnTestCaseStart` property will be something like this:

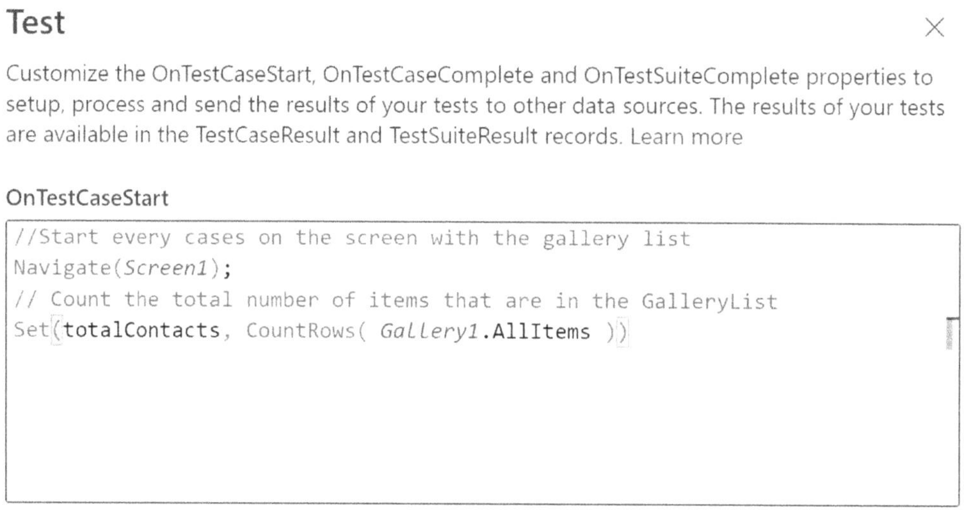

Test   ×

Customize the OnTestCaseStart, OnTestCaseComplete and OnTestSuiteComplete properties to setup, process and send the results of your tests to other data sources. The results of your tests are available in the TestCaseResult and TestSuiteResult records. Learn more

OnTestCaseStart

```
//Start every cases on the screen with the gallery list
Navigate(Screen1);
// Count the total number of items that are in the GalleryList
Set(totalContacts, CountRows( Gallery1.AllItems ))
```

Figure 5.16 – OnTestCaseStart

Coming back to our steps in the test case, namely the **Check Email in row 5** step, we will add a new assertion to our second test where we will check that the number of elements in `GalleryList` corresponds to the value we initialized before starting the tests:

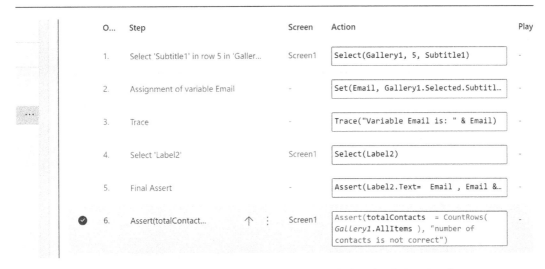

O...	Step	Screen	Action	Play
1.	Select 'Subtitle1' in row 5 in 'Galler...	Screen1	`Select(Gallery1, 5, Subtitle1)`	-
2.	Assignment of variable Email	-	`Set(Email, Gallery1.Selected.Subtitl...`	-
3.	Trace	-	`Trace("Variable Email is: " & Email)`	-
4.	Select 'Label2'	Screen1	`Select(Label2)`	-
5.	Final Assert	-	`Assert(Label2.Text= Email , Email &...`	-
✓ 6.	Assert(totalContact... ↑ ⋮	Screen1	`Assert(totalContacts = CountRows( Gallery1.AllItems ), "number of contacts is not correct")`	-

Figure 5.17 – Assertion to check that a variable was declared in OnTestCaseStart

Now, we can play back our tests to check the result, as shown here:

Figure 5.18 – Playing back tests with OnTestCaseStart

Now, let's force an error in our tests to check how it is detected by Test Studio.

By playing with the `totalContacts` variable (by increasing, for instance, the number to assert artificially), *we can force an error in the final assertion.*

The new assertion will check if the number of rows will be equal to `totalContacts variable` plus one. We know that this should not be correct:

Figure 5.19 – Forcing an error with a final assertion

As a result, an error will be displayed when we play back our tests:

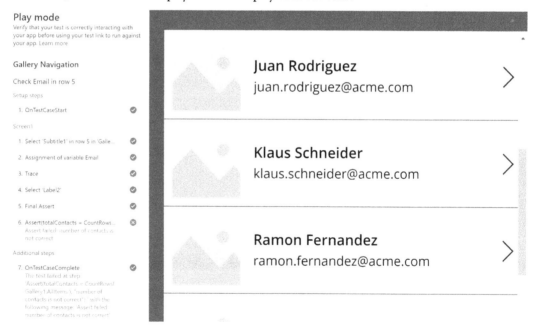

Figure 5.20 – An error is displayed in the playback

In the `OnTestCaseStart` event, we have access to the `TestCaseInfo` record, which contains details for the current test that is executing, including properties such as the test case's name, description, and ID, as well as the name, description, and ID of the test suite that the case belongs to.

Processing and saving test results

When running tests through a browser, you cannot view the test panel that is normally displayed in Test Studio. This can make it difficult to monitor the progress of your tests or determine whether a test has passed or failed.

For instance, by using the play link feature, you can duplicate a link so that you can execute a test in an independent browser beyond Test Studio's environment. This can be beneficial for integrating your tests into a continuous integration and deployment workflow, such as Azure DevOps. To capture test results beyond Test Studio, you can use the `OnTestCaseComplete` and `OnTestSuiteComplete` events to manage and send your test results to multiple data repositories or platforms. These properties are available in the test object and are triggered when a test case or suite is completed.

In these events, we can leverage the `TestCaseResult` and `TestSuiteResult` records to access information about specific test cases and the entire test suite. All the properties included in these collections can be found here: `https://learn.microsoft.com/en-us/power-apps/maker/canvas-apps/working-with-test-studio#processing-test-results`.

The `OnTestCaseComplete` and `OnTestSuiteComplete` properties can be combined with these records so that they can be sent to or integrated with different data sources/services for storage, further processing, or notifications (such as integrations with Power Automate, persisting in Dataverse, and so on).

Also, these properties can be customized to trigger when a test case or suite completes and can be utilized in continuous build and release pipelines, such as Azure DevOps, to determine whether to proceed with app deployment.

> **Note**
>
> For our example, we will save our results to an Excel file with the structure needed as a variation of the example in the official documentation, but the same code can be applied if you wish to save these results in Dataverse. You can check the version with Dataverse here: `https://learn.microsoft.com/en-us/power-apps/maker/canvas-apps/working-with-test-studio#processing-test-results`.

Saving results to an Excel file

First, we will generate an *Excel file with the required columns*, as shown in the following screenshot. These columns can also be checked in the code that we will introduce to save the test results:

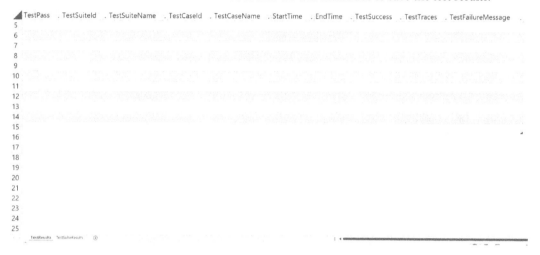

Figure 5.21 – Excel TestResults sheet structure

> **Note**
>
> The Excel sheet we've created is available in this book's GitHub repository, along with the base gallery application.

For the TestResults sheet, we will use the TestPass, TestSuiteId, TestSuiteName, TestCaseId, TestCaseName, StartTime, EndTime, TestSuccess, TestTraces, and TestFailureMessage columns.

For the TestSuiteResults sheet, we will use the TestSuiteID, TestSuiteName, StartTime, EndTime, TestPassCount, and TestFailCount columns:

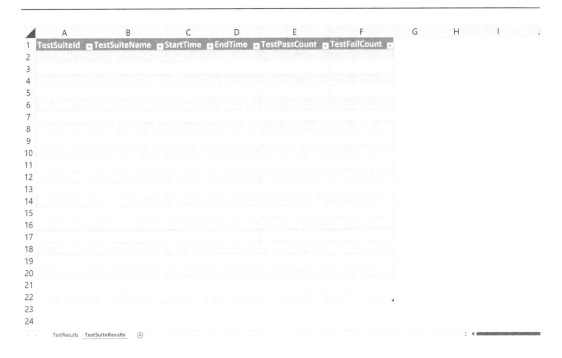

Figure 5.22 – Excel TestSuiteResults sheet structure

Each sheet should be formatted as a table, and we must assign the name of the table that we want to use later in our code – for example, `AppTestResults` and `AppTestSuiteResults`:

Figure 5.23 – Excel table names

Once we have completed the Excel template, we can upload this file to a OneDrive account that we will access later through a connector via our Power App:

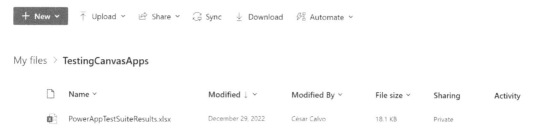

Figure 5.24 – OneDrive for Enterprise folder containing an Excel file

Using the **OneDrive for Enterprise connector**, we will add a connection to our Excel file by selecting the two tables that we have created for each sheet:

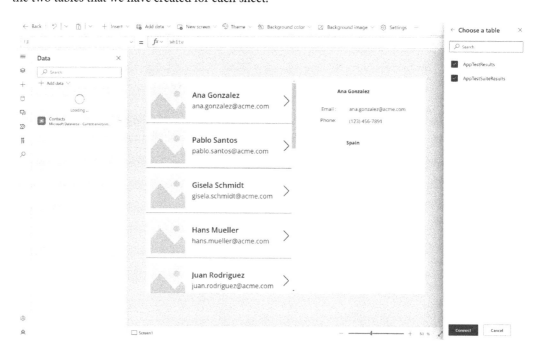

Figure 5.25 – OneDrive for Enterprise folder connection to Excel

The result will be as follows on our data screen:

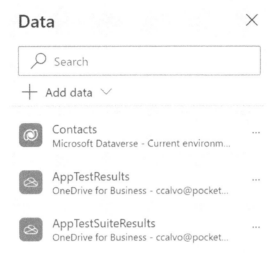

Figure 5.26 – Connections to Excel tables in OneDrive

We already have the necessary connection to save our results in the Excel file that we created previously. We must save and publish our new version of the application, and then return to our Test Studio screen.

We will apply this code in the OnTestCaseComplete step:

```
//Save to Excel
Patch(AppTestResults
, Defaults(AppTestResults)
, {       TestPass: "Name of the test:" &
           TestCaseResult.TestCaseName & ":" & Text(Now())
         ,TestSuiteId:   TestCaseResult.TestSuiteId
         ,TestSuiteName: TestCaseResult.TestSuiteName
         ,TestCaseId: TestCaseResult.TestCaseId
         ,TestCaseName: TestCaseResult.TestCaseName
         ,StartTime: TestCaseResult.StartTime
         ,EndTime: TestCaseResult.EndTime
         ,TestSuccess: TestCaseResult.Success
         ,TestTraces: JSON(TestCaseResult.Traces)
         ,TestFailureMessage: "Error: " &
           TestCaseResult.TestFailureMessage
}
);
```

To do this, click the **Test** button or **Results Test applied - View**, as we did previously:

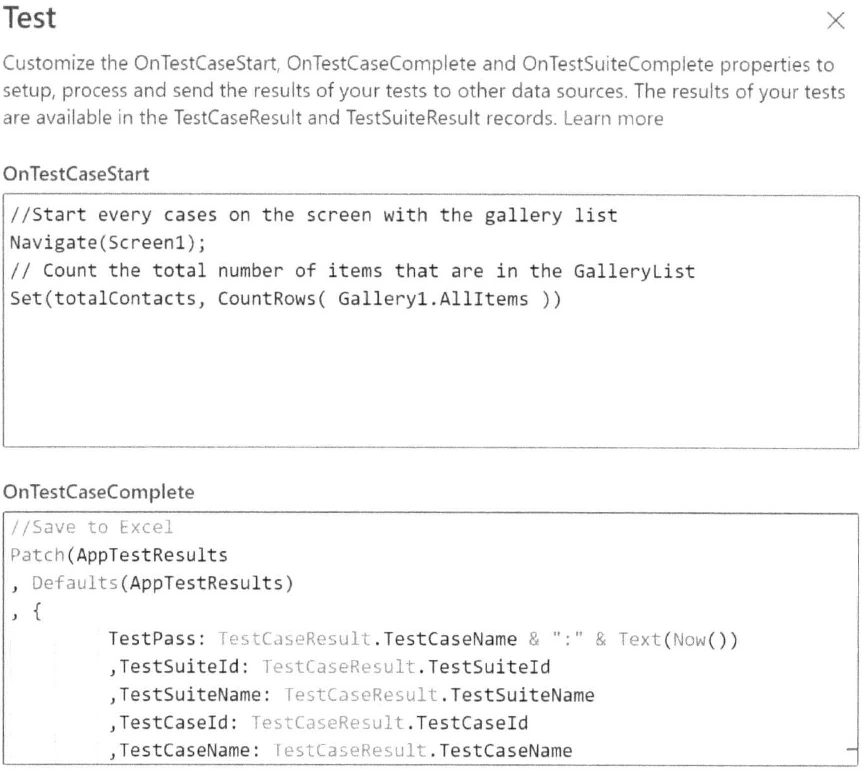

Test   ✕

Customize the OnTestCaseStart, OnTestCaseComplete and OnTestSuiteComplete properties to setup, process and send the results of your tests to other data sources. The results of your tests are available in the TestCaseResult and TestSuiteResult records. Learn more

OnTestCaseStart

```
//Start every cases on the screen with the gallery list
Navigate(Screen1);
// Count the total number of items that are in the GalleryList
Set(totalContacts, CountRows( Gallery1.AllItems ))
```

OnTestCaseComplete

```
//Save to Excel
Patch(AppTestResults
, Defaults(AppTestResults)
, {
        TestPass: TestCaseResult.TestCaseName & ":" & Text(Now())
        ,TestSuiteId: TestCaseResult.TestSuiteId
        ,TestSuiteName: TestCaseResult.TestSuiteName
        ,TestCaseId: TestCaseResult.TestCaseId
        ,TestCaseName: TestCaseResult.TestCaseName
```

Figure 5.27 – OnTestCaseComplete

Finally, we will apply this code in the OnTestSuiteComplete step:

```
//Save to Excel
Patch(AppTestSuiteResults
    , Defaults(AppTestSuiteResults)
    , {
        TestSuiteId: TestSuiteResult.TestSuiteId
        ,TestSuiteName: TestSuiteResult.TestSuiteName
        ,StartTime: TestSuiteResult.StartTime
        ,EndTime: TestSuiteResult.EndTime
        ,TestPassCount: "Number of tests passed: " &
          TestSuiteResult.TestsPassed
        ,TestFailCount: "Number of tests failed: " &
          TestSuiteResult.TestsFailed
    }
);
```

Our `OnTestSuiteComplete` step will look something like this:

OnTestSuiteComplete

```
//Save to Excel
Patch(AppTestSuiteResults
    , Defaults(AppTestSuiteResults)
    , {
        TestSuiteId: TestSuiteResult.TestSuiteId
        ,TestSuiteName: TestSuiteResult.TestSuiteName
        ,StartTime: TestSuiteResult.StartTime
        ,EndTime: TestSuiteResult.EndTime
        ,TestPassCount: TestSuiteResult.TestsPassed
```

Figure 5.28 – OnTestSuiteComplete

Now, we can save the changes and run the tests. Finally, we will be able to access the results and traces of the tests in the Excel file via OneDrive:

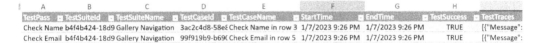

TestPass	TestSuiteId	TestSuiteName	TestCaseId	TestCaseName	StartTime	EndTime	TestSuccess	TestTraces
Check Name	b4f4b424-18d9	Gallery Navigation	3ac2c4d8-58e8	Check Name in row 3	1/7/2023 9:26 PM	1/7/2023 9:26 PM	TRUE	[{"Message":
Check Email	b4f4b424-18d9	Gallery Navigation	99f919b9-b69C	Check Email in row 5	1/7/2023 9:26 PM	1/7/2023 9:26 PM	TRUE	[{"Message":

Figure 5.29 – Test results in the Excel file

In addition to storing the tests in Excel, using the same code, we can store the tests in Dataverse tables with these columns. We also have the following options:

- Send the results to a flow in Power Automate that receives the JSON with the results and stores them in a test plan in Azure DevOps:

```
SaveTestResultstoTestPlan.Run(JSON(TestCaseResult))
```

- Email the results (that is, send an email with the link to the OneDrive file that was created previously:

```
// Send the email with the link. Office365Outlook is a connector
in // our Power App
Office365Outlook.SendEmailV2("someone@packt.com", "Test case
results", "Please find the link to the OneDrive file containing
the test results: https://example.onedrive.com/link/to/file ");
```

- Receive an app notification of the test result.

  For example, we can receive a notification after the test is completed when playing the test in a browser, outside of Test Studio. In this case, the code should be in the `OnTestCaseComplete` event, as shown in the following screenshot:

OnTestCaseComplete

```
Notify("Test  notification: " & TestCaseResult.TestCaseName & " : "
       & If( TestCaseResult.Success, " Test Passed", "Test Failed " &
TestCaseResult.TestFailureMessage)
       ,If( TestCaseResult.Success, NotificationType.Success,
NotificationType.Error)
)
```

Figure 5.30 – An app notification showing the test result

Note that these last two actions can also be included in a Power Automate flow, and in this way, we can use all the out-of-the-box connectors available in Power Automate, and even custom connectors, to send this information to any required external quality/testing system.

Summary

In this chapter, we explored the main concepts of the Test Studio tool, which is currently the standard for testing the automation of canvas apps. Using this tool, we reviewed the most relevant concepts related to testing automation, such as test suites, test cases, test assertions, and the limitations that may arise when using them.

We demonstrated the steps required to establish the required environment for developing the examples outlined in this book. Then, we applied this environment to a simple example that involved using a contact gallery to record, customize, and perform a playback in Test Studio.

Finally, we demonstrated how the results of these tests can be saved in a database or file and also sent for processing in external tools and monitored in deployment processes. In the next chapter, we will explore the anticipated evolution of this tool toward a new framework called Test Engine, which will address some of the limitations we have observed in this chapter.

6
Overview of Test Engine, Evolution, and Comparison

Test Engine serves as an innovative testing platform for Power Apps, designed to offer a comprehensive testing solution for all Power Apps and simplify the incorporation of automated testing within app development workflows. It is *an open source initiative hosted on GitHub and backed by Microsoft*.

Test Engine is built on the Playwright browser testing platform, and the **Power Fx** language is utilized for defining tests in **YAML** files, which are simple and comprehensive. We will explore both elements in more detail later.

We are facing the initial preview of Test Engine, which is an evolution in the testing tools for Power Apps, considering the most common use cases but taking a new approach with the use of the **Playwright** testing platform and adopting an open source philosophy.

This initial version only supports testing on Power Apps canvas applications, although work is already underway to include support for model-driven apps and integrations with Azure DevOps and **continuous integration and continuous development** (**CI/CD**) processes in general. As an example, at the time of writing this book, the integration of Test Engine with the **Power Platform Command-Line Interface** (**PAC CLI**) has been announced in public preview, and we will cover this integration in *Chapter 7*.

Among the advantages of Test Engine that we will describe in the following sections is the ability to define *mock responses for connectors*, and this will enable creators to test Power Apps separately from the external APIs that the app is connected to, creating a full unit test.

Test Engine also includes a `Screenshot` function and the option to *record a video* of the entire test run. These can prove to be advantageous for diagnosing and examining unsuccessful tests since they enable the test creator to observe precisely what the end user perceives while the test is executed.

Another advantage of Test Engine is that tests are written using Power Fx and *referencing the names of controls* is defined at design time, so test plans only need to be updated if changes are made to the app itself. This contrasts with traditional testing methods that require the test author to interact with the browser **Document Object Model** (**DOM**) of the app as it is rendered in the web player, which can be difficult to get right and frequently breaks.

> **Note**
>
> As previously mentioned, Test Engine is in an experimental phase at the time of writing this book, but plans are in place to add support for all types of Power Apps and improve the tooling to facilitate integration with CI/CD systems such as GitHub and Azure DevOps.
>
> At the time of writing this book, the *integration with PAC CLI is in public preview*. It presents an alternative method of working with Test Engine. We will delve into this option in *Chapter 7*.

This chapter consists of the following topics:

- *Getting to know Test Engine*: This section explores the concept and key components of Test Engine
- *Understanding the benefits of Test Engine*: Here, we analyze the main benefits of leveraging Test Engine
- *Knowing the limitations of Test Engine*: This part covers the current limitations of Test Engine
- *Prerequisites and build process of Test Engine*: This section describes what we need to install as a prerequisite and the steps required to obtain the Test Engine binary from the repository code

Technical requirements

To follow the examples in this book, you must download the corresponding files from the designated GitHub repositories. We will utilize two repositories, as follows:

1. The example and code repository provided by Microsoft for the Test Engine component, which will be discussed in the next chapter.

   Link: `https://github.com/microsoft/PowerApps-TestEngine`

2. The GitHub repository specifically created for this book.

   Link: `https://github.com/PacktPublishing/Automate-Testing-for-Power-Apps`

Access to a functioning Power Platform environment is necessary in order to install the provided examples or create new ones. The Power Platform developer environment appears to be the most suitable solution for this purpose. We have already described how to access it in the previous chapter.

We will need to install the **.NET Core 6.0.x software development kit** (SDK) and add it to the environment variables of our operating system. We will cover the details on how to do this in a later section of this chapter, *Prerequisites and build process of Test Engine*.

Another requirement for using Test Engine is the installation of **PowerShell**, as Test Engine uses it to install the necessary browsers. To keep all the instructions together, the entire process will be described in the aforementioned section.

While not strictly necessary for this chapter, installing an **integrated development environment** (IDE) such as Visual Studio Code can be beneficial, especially as we are working with and editing YAML files.

Getting to know Test Engine

Let's begin exploring the elements that form the foundation of Test Engine. We'll start by discussing the ability to create tests using the Power Fx language, which we reviewed in *Chapter 3* of this book.

Power Fx test authoring

Power Fx is a low-code language for Power Platform and it is also used for Test Engine.

Defining your tests doesn't require programming languages such as C# or JavaScript. Instead, tests are specified using straightforward YAML files that incorporate the well-known Power Fx language. Familiar Power Fx functions, such as `Select`, `Assert`, `SetProperty`, and `Index`—which you might recognize from **Test Studio**—can be used to outline the steps in your test case.

Here's an example of a Test Engine test step using these functions:

```
testSteps: |
  =
  Select(Gallery1, CountRows( Gallery1.AllItems ), NextArrow1);
  Screenshot("galleryadd_end.png");
```

Figure 6.1 – Power Fx syntax for test steps

Any documented Power Fx functions can be used within Test Engine. In addition to the general Power Fx functions, there are several specifically defined functions for the test framework. These include the following:

- `Assert`
- `Screenshot`
- `Select`
- `Wait`

Let's take a quick look at each of these:

- `Assert`:

  - `Assert` (Boolean expression)

  - `Assert` (Boolean expression, message)

  - The `Assert` function takes in a Power Fx expression that should evaluate to a Boolean value. If the value returned is `false`, the test will fail.

  Example:

  ```
  Assert(Label1.Text = "1");
  Assert(Label1.Text = "1", "Checking that the Label1 text is set
  to 1");
  ```

- `Screenshot` (filename of screenshot):

  - This function will capture a screenshot of the app at the current point in time. The screenshot file will be saved to the test output folder with the name provided.

> **Note**
> Only `jpeg` and `png` files are supported. Recording videos can be enabled in a parameter of `TestSettings`, as we will check later.

  Example:

  ```
  Screenshot("buttonClicked.png")
  ```

- `Select`:

  - `Select` (control)

  - `Select` (control, row, or column)

  - `Select` (control, row, or column, child control)

  - `Select(Index(gallerycontrol.AllItems, row or column).child control)`

  - This function is identical to the `Select` function in Power Apps

  Example:

  ```
  Select(Button1)
  Select(Gallery1,1)
  Select(Gallery1,1,Button1)
  Select(Index(Gallery1.AllItems, 2).Icon2)
  ```

```
Select(Index(Index(Gallery1.AllItems, 1).Gallery2.AllItems,
4).Icon3);
```

- `Wait`:

    - `Wait` (control, property, value)

    - This function will wait for the property of the control to equal the specified value

    Example:
    ```
    Wait(Button1, DisplayMode, Edit)
    Wait(Gallery1, Visible, false)
    ```

Now that we've covered how to use Power Fx syntax to complete the steps of our tests, let's take a look at how test definitions are created using the industry standard YAML language, which is commonly used in DevOps operations.

YAML format

Power Apps Test Engine uses YAML to define tests, following the same guidelines as Power Fx. This allows for a consistent and familiar syntax for users already familiar with Power Fx.

YAML is an industry-standard language with well-established grammar and many editors, tools, and libraries for working with it. For instance, YAML is commonly used in configuring CI/CD pipelines in Azure DevOps and GitLab.

We will delve deeper into the topic throughout the book, but for now, let's provide some information about the YAML schema definition for Power Apps Test Engine.

Testing the YAML schema definition

A YAML schema is used to define a single test and includes properties such as the name and description of the test suite, the persona of the user performing the test, and the logical name or ID of the app being launched. In addition, it includes options for defining network request mocks, test cases, and test suite triggers.

The following screenshot shows an overview of a typical YAML schema:

```
C: > te3 > PowerApps-TestEngine > samples > basicgallery > ! canvascomponent.fx.yaml > {} testSuite
 1    testSuite:
 2      testSuiteName: CanvasComponent
 3      testSuiteDescription: ''
 4      persona: User1
 5      appLogicalName: cccatp_chat_06d07
 6      appId: ''
 7      onTestCaseStart: ""
 8      onTestCaseComplete: ""
 9      onTestSuiteComplete: ""
10      networkRequestMocks:
11      testCases:
12      - testCaseName: Check Chat button
13        testCaseDescription: ''
14        testSteps: |
15          =
16          Screenshot("UserCard_loaded.png");
17          SetProperty(UserCard_1.Status, "Test");
18          Assert(UserCard_1.Status = "Test", "Make sure status is set to Test");
19          Screenshot("Usercard_statuschanged.png");
20          SetProperty(UserCard_1.UserRole, "Developer");
21          Assert(UserCard_1.UserRole = "Developer", "Make sure role is set to Developer");
22          Screenshot("Usercard_end.png");
23
24    testSettings:
25      locale: "en-US"
26      recordVideo: true
27      headless: false
28      browserConfigurations:
29        - browser: Chromium
30    environmentVariables:
31      users:
32        - personaName: User1
33          emailKey: user1Email
34          passwordKey: user1Password
```

Figure 6.2 – YAML structure for Test Engine

In the YAML schema, we will encounter some properties *required* for the test definition, such as the following:

- `testSuiteName`: This is the name of the test suite.

- `persona`: This is the user that will be logged in to perform the test. This must match a persona listed in the `users` section.

- `appLogicalName`: This is the logical name of the app that is to be launched. It can be obtained from the solution. For canvas apps, you need to add it to a solution to obtain it.
- `testCases`: Defines test cases in the test suite. The test scenarios within the test suites are executed consecutively, and the app's condition is preserved throughout all test cases present in a suite. We will add our Power Fx code in the `testSteps` property included in this section, as we will check in more detail in *Chapter 7*.

The following properties are *optional*:

- `testSuiteDescription`: Additional information describing what the test suite does.
- `appId`: This is the ID of the app that is to be launched. This is required and used only when the app's logical name is not present. `appId` should be used only for canvas apps that are not in the solution.
- `networkRequestMocks`: Defines network request mocks needed for the test. We will delve into this functionality in *Chapter 11*.
- `onTestCaseStart`: Defines the steps that need to be triggered for every test case in a suite before the case begins executing.
- `onTestCaseComplete`: Defines the steps that need to be triggered for every test case in a suite after the case finishes executing.
- `onTestSuiteComplete`: Defines the steps that need to be triggered after the suite finishes executing.

The `testSettings` feature in the test plan defines parameters such as browser configurations, video recording preferences, and headless browser mode. We will adjust some of these parameters in *Chapter 7*.

References to user credentials can be found under the `environmentVariables` section, represented as a list of users. Each user is defined with a `personaName` property, an `emailKey` property, and a `passwordKey` property. It's important to note that storing credentials directly in test plan files is not supported. This necessitates the creation of variables in our environment to store these values—a process we will explore in detail later.

One of the major benefits of the new Test Engine platform is the ability to *reuse tests created with Test Studio*. In the next section, we'll explore how to take advantage of this feature.

Harnessing Test Studio's recorded tests with Test Engine

One of the features of Test Engine is *the ability to download recorded tests from Test Studio*. This allows users to reuse previously recorded tests in Test Engine, making the testing process more efficient.

To obtain tests from Test Studio, users can click the **Download suite** button in Test Studio to retrieve the test plan. If there are multiple test suites, users must select the test suite they wish to download. The process is illustrated in the following screenshot:

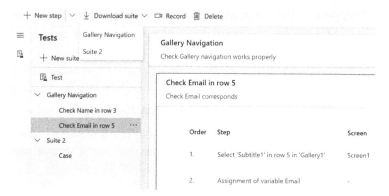

Figure 6.3 – Downloading a test plan from Test Studio

Additionally, there is also a **Download** button available under each test suite, which allows users to download individual test suites:

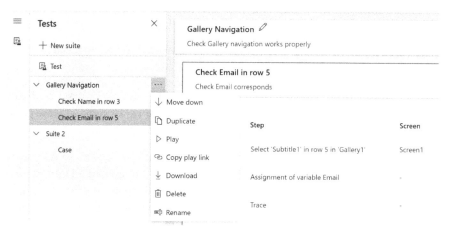

Figure 6.4 – Downloading test suites from Test Studio

> **Note**
>
> In preparation for reusing these tests with Test Engine, users must create a configuration file and user credentials. We will delve into this in the next chapter.

An important aspect of Test Engine is the underlying technology it uses: Playwright. While this book is not intended to cover Playwright in depth, let's highlight some details in the next section.

Playwright

Playwright is an automation and web testing tool created by Microsoft, and Test Engine translates the Power Fx syntax into Playwright tests. It allows for the automation of user interactions within the app, simulating the actions of a real user.

One of the main advantages of Playwright is its ability to run tests in *headless mode*. This means that the browser is launched without a GUI, allowing tests to run faster and more efficiently. This is *particularly important in the context of CI/CD*, where tests are typically run on a regular basis and are integrated into the development process. By using headless mode, tests can be completed more quickly, reducing the overall time required for the CI/CD pipeline to complete.

Furthermore, when testing in headless mode, Playright allows for *testing in parallel*, which can speed up the test execution time, allowing for faster feedback on the development process and helping to increase the overall test coverage.

Additionally, Playwright offers a wide range of capabilities, including the ability to click, type, and navigate, as well as advanced functionality such as file uploads and form submissions.

Another key feature of Playwright is its *support for multiple contexts*. Context refers to a browsing session, and Playwright allows for the creation of multiple contexts within a single browser instance. This allows for the simultaneous execution of multiple tests or the ability to switch between different contexts within a test.

To learn more about Playwright, we recommend visiting the official website: `https://playwright.dev/`. This website provides detailed documentation, tutorials, and examples to help you get started with Playwright.

Let's refocus in the next section on Test Engine and provide some notes about how this framework is expected to evolve.

Evolution of Test Engine

As previously mentioned, the initial release of Test Engine focuses on providing the ability to author tests for Power Apps canvas applications. However, it is important to note that the development team has plans to continue iterating on this project and expanding its capabilities. This includes adding support for all types of Power Apps, as well as enhancing the tooling to make it easier to integrate with popular CI/CD systems such as GitHub and Azure DevOps.

To stay informed about the latest updates and new features, it is recommended to regularly visit the GitHub page of the project at the following link: `https://github.com/microsoft/PowerApps-TestEngine`. This is where the development team shares information about new releases, upcoming features, and other important updates.

Given that Test Engine is now integrated within *PAC CLI*, it is worthwhile referring to the Power Apps release planner for any related information or updates: `https://releaseplans.microsoft.com/en-US/?app=Power+Apps`.

In the next section, we are going to describe the *additional advantages* that the new Test Engine platform for Power Apps presents.

Understanding the benefits of Test Engine

Among the many benefits, we will highlight the following:

- Making mock responses for connectors
- The `Screenshot` function
- Video recording

We will also see a short comparison between the different current possibilities of performing interface test automation in Power Apps, comparing Test Studio with Test Engine and Playwright, which is the engine on which Test Engine is based.

Connector mocking

Test Engine permits defining *mock responses for connectors*, enabling the evaluation of Power Apps in isolation from the external APIs it connects with.

This functionality can be advantageous while examining applications that access endpoints with side effects, allowing full unit testing—for example, calling an API to perform some operations in certain records. The mock responses are defined using the `NetworkRequestMocks` property, which includes the following fields: `requestURL`, `responseDataFile`, `Method`, `Headers`, and `requestBodyFile`, as we saw before when we were describing the YAML structure.

> **Note**
>
> This functionality will be covered in more detail in *Chapter 11*. However, it's important to note that, while this functionality is very helpful, a real integration test will be necessary in the end to verify the response in an actual environment.

Screenshot function and video recording

One of the most notable features of Test Engine is the `Screenshot` function, as shown in *Figure 6.5*. This function enables capturing a screenshot of an app's condition at any moment during a test's execution. This can be incredibly useful for capturing what the end user sees during different stages of a test suite. For instance, a screenshot can be taken at the start of each test case and after the completion of the test suite and can be saved in `.jpeg` or `.png` file formats.

A video recording can be configured in the `testSettings` parameter, where we will find the `recordVideo` parameter, also shown in *Figure 6.5*. Simply setting this parameter to `true` will enable the recording of our test execution.

These screenshots and videos can be crucial for validating the correct test results, especially if tests are executed in headless mode:

```
testCases:
- testCaseName: Check Chat button
  testCaseDescription: ''
  testSteps: |
    =
    Screenshot("UserCard_loaded.png");
    SetProperty(UserCard_1.Status, "Test");
    Assert(UserCard_1.Status = "Test", "Make sure status is set to Test");
    Screenshot("Usercard_statuschanged.png");
    SetProperty(UserCard_1.UserRole, "Developer");
    Assert(UserCard_1.UserRole = "Developer", "Make sure role is set to Developer");
    Screenshot("Usercard_end.png");

testSettings:
  locale: "en-US"
  recordVideo: true
  headless: false
  browserConfigurations:
    - browser: Chromium
```

Figure 6.5 – Screenshot function and recordVideo parameter

Now that we have explored the features of Test Studio in the previous chapter and discussed Test Engine in this chapter, as well as having touched on the underlying Playwright technology, let's take a moment to compare these approaches and discuss when each one would be the best fit.

Comparison with Test Studio and Playwright

First of all, we have to mention that Test Engine is a project that is still in the experimental phase. It is a project created to *expand the functionality present in Test Studio*, but in its current phase, it lacks what we could consider a low-code approach.

For example, the instructions to follow are more typical of a developer tool, and it lacks a GUI interface that allows for more user-friendly operations, although the functionality of downloading tests recorded with Test Studio greatly facilitates this step.

On the other hand, Test Studio is the tool currently recommended by Microsoft for the development of interface tests in Power Apps, and it is at a fully mature stage. It is *fully integrated into the Power Platform environment* and is a tool fully designed to *create tests visually* with the operation recorder. It also has supported integration with Azure DevOps scenarios.

However, there are *certain limitations in Test Studio* that make us look forward to the evolution of Test Engine (tests are stored within the app; lack of collaboration mode and extensibility…). For example, Test Studio is only limited to canvas apps, and although Test Engine is currently only compatible with canvas apps, the roadmap includes coverage of model-driven apps, as illustrated in *Figure 6.6*.

There are other series of limitations that are *covered by Test Engine*, such as the following:

- Testing of components within the application
- Testing of components developed with the **Power Apps Component Framework** (**PCF**)
- Nested galleries
- Lack of mock responses for connections
- `Screenshot` and video recording functions

However, *both tools use the Power FX syntax* for test authoring, and assuming that the evolution of Test Engine is to develop a more user-friendly interface, both tools can be considered within the scope of low code.

This is not the case for the framework that we include in this comparison, which is Playwright. When talking about Test Engine, we have described how it is based on Playwright, and we have talked about the characteristics of this test engine developed by Microsoft.

However, despite the power of this engine, it is important to keep in mind that it is a more generic engine and *requires the development of tests in JavaScript or TypeScript*, which would not allow us to consider it as a low-code framework.

Therefore, we are talking about a tool that allows for a *higher level of control* and that allows for testing more advanced scenarios and any type of application that runs on a browser, but *without the advantage of being able to use the Power FX syntax and without the visual environments* provided by Power Platform (although it does offer a test recorder for added convenience):

```
test('create a contact', async ({ page }) => {
    // Open the contacts area from the sitemap
    await page.goto(config.appUrl);
    await page.waitForNavigation(),

    // Click on the New button in the command bar
    await page.click('[aria-label="New"]')
    await page.waitForNavigation(),

    // Complete some details on the contact form
    await page.fill('[aria-label="First Name"]', 'Test');
    await page.fill('[aria-label="Last Name"]', `Contact (${Date.now()})`);

    // Click on the Save button in the command bar
    await page.click('[aria-label="Save (CTRL+S)"]')
    await page.waitForNavigation(),

    // Check if the next form contains the contacts ID
    await page.waitForFunction('Xrm.Utility.getPageContext().input.entityId !== null');
});
```

Figure 6.6 – Playwright test: Creating a contact in a model-driven app

To complement this comparison, let's take a look at the current limitations presented by Test Engine.

Knowing the limitations of Test Engine

As previously mentioned, Test Engine is an experimental project and still has a number of limitations.

While work is ongoing to provide full control coverage, *support for charts, media, and timer and mixed reality (MR) controls is currently unavailable*. However, as this technology is still in the experimental phase, it is possible that changes may occur in the future. Please keep this in mind when considering the information provided.

Additionally, at the current stage, *Test Engine does not support child controls within components*, making it difficult to test certain aspects of an application.

It is also important to note that *multi-factor authentication (MFA) is not supported*. This means that in order to run tests, users must use an account that requires only a username and password. This is a limitation that limits the current usage for development or testing environments in enterprise organizations.

> **Note**
> As of writing, there is no available information on whether MFA support is planned for the Power Apps Test Engine roadmap. However, it is important to keep an eye on updates and announcements from Microsoft.

In the following section, we will identify the elements required to build our Test Engine binary.

Prerequisites and build process of Test Engine

In the next chapter, we will guide you through the process of using Test Engine with the app developed in *Chapter 5* or with the sample projects available for download on GitHub. Before diving into the testing process, however, it's important to ensure that all necessary prerequisites are met and that the executable is properly generated.

To begin, we will start by either cloning or downloading the project from GitHub. From there, we will walk you through the process of checking for prerequisites and generating the executable, before diving into the testing process itself.

> **Note**
>
> We assume that you already have cloned or downloaded the Power Apps Test Engine repository (`https://github.com/microsoft/PowerApps-TestEngine`) from *Chapter 5*.

To use the Test Engine binary, we need to compile the code contained in this repository. We will describe how to do this in the next sections. While we've already mentioned the alternative of using the new version of PAC CLI, we believe it's important to describe how to use it directly from the repository. This provides the possibility of extending it, as we will illustrate in *Chapter 11*.

Installing the prerequisites to compile Test Engine

Before building the Power Apps Test Engine executable, it's important to ensure that the following prerequisites are met:

- **.NET Core 6.0.x SDK**: Test Engine is built using .NET Core, so you will need to have the .NET Core 6.0.x SDK installed on your computer. You can download the SDK from the official .NET website (`https://dotnet.microsoft.com/download/dotnet-core`).

- **MSBuildSDKsPath environment variable**: Test Engine requires the `MSBuildSDKsPath` environment variable to be set to the .NET Core 6.0.x SDK. This variable tells the build system where to find the necessary build tools and libraries. To set the variable, follow these steps:

  I.   Open the **Start** menu and type `Environment Variables` into the search bar.

> **Note**
>
> There are some minor differences between the user interfaces of Windows 10 and Windows 11, but the basic process of editing environment variables is the same. The process described here is based on Windows 11, so please take into account any minor differences you may observe if you are working with Windows 10.

II.   Click on the **Edit the system environment variables** button. This will then take you to the following screen:

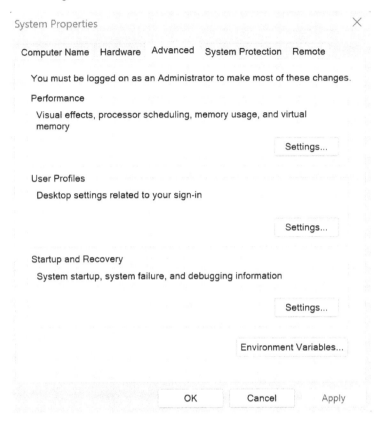

Figure 6.7 – Editing system environment variables

III. In the **System Properties** window, click on the **Environment Variables...** button.

IV.   In the **System variables** section, scroll down and find the MSBuildSDKsPath variable:

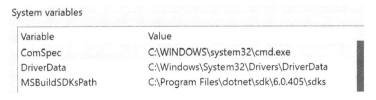

Figure 6.8 – MSBuildSDKsPath variable

V.   If the variable exists, edit it to point to the .NET Core 6.0.x SDK path. If the variable does not exist, click the **New** button and add the variable with the correct path.

> **Note**
>
> Refer to the following section to verify the accurate path of the .NET SDK. Additionally, take note of the \sdks subfolder in the corresponding environment variable, which should match the actual subfolder. Refer to *Figure 6.9.*

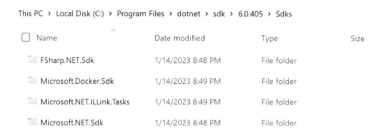

Figure 6.9 – .NET SDK folders

- **PowerShell**: Test Engine uses PowerShell scripts to install the required browsers. *Make sure that you have PowerShell installed on your computer.* If you don't, you can download it from the official PowerShell website (https://docs.microsoft.com/en-us/powershell/scripting/install/installing-powershell). Ensure that you have the appropriate permissions required to run PowerShell scripts on your machine.

Once you have these prerequisites installed, you should be ready to build the Test Engine platform using the instructions provided in the repository. If you encounter any problems related to the correct setting of the .NET SDK path, let's take a moment to double-check that it is correctly assigned in the next section, especially if you are experiencing issues when compiling the binary.

The path of the .NET SDK

The correct path to set the MSBuildSDKsPath environment variable to the .NET Core 6.0.x SDK depends on your operating system and the location of the SDK on your machine.

On Windows, the default location for the .NET Core 6.0.x SDK is typically C:\Program Files\dotnet\sdk\6.0.x\.

> **Note**
>
> You can also check the installation location of the .NET Core 6.0.x SDK by running the following command in Command Prompt or a PowerShell window: dotnet --info.

This will display the installed .NET Core versions and their respective installation paths.

Once you have determined the correct location of the .NET Core 6.0.x SDK on your machine, you can set the MSBuildSDKsPath environment variable to this path.

In addition to the method previously described, on Windows, you can set the environment variable in Command Prompt or a PowerShell window by using the `setx` command (`https://learn.microsoft.com/en-us/windows-server/administration/windows-commands/setx`), as follows:

```
setx MSBuildSDKsPath "C:\Program Files\dotnet\sdk\6.0.x\sdks"
```

Refer to the following screenshot to see how to apply this command in PowerShell:

Figure 6.10 – Setting MSBuildSDKsPath variable via PowerShell

> **Note**
> Take note of the `\sdks` subfolder in the corresponding environment variable, which should match the actual subfolder.

You may need to restart your terminal or log out and log back in for the environment variable to take effect.

Creating the executable

Once you have the repository cloned or downloaded, and with the prerequisites previously described fulfilled, you can build it by following these instructions:

1. Open a terminal window and navigate to the directory where you have downloaded the project and the `PowerAppsTestEngine` folder inside the `src` folder named `\src\PowerAppsTestEngine`, as illustrated here:

Name	Date modified	Type	Size
config.json	12/13/2022 6:55 AM	JSON Source File	1 KB
InputOptions.cs	12/13/2022 6:55 AM	C# Source File	1 KB
PowerAppsTestEngine.csproj	12/13/2022 6:55 AM	C# Project Source File	1 KB
Program.cs	12/13/2022 6:55 AM	C# Source File	6 KB

> USB Drive (D:) > PowerApps-TestEngine-main > src > PowerAppsTestEngine

Figure 6.11 – PowerAppsTestEngine folder

In *Figure 6.12*, you can see how to change to the previously mentioned folder. Please adapt this command to the drive and path that you are using on your operating system:

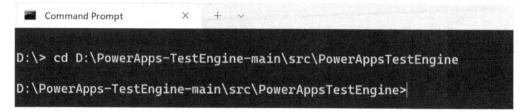

Figure 6.12 – Changing to the PowerAppsTestEngine folder in Command Prompt

- Run the following command to build the project:

```
dotnet build
```

After running this command, we should obtain something like this:

```
D:\>cd D:\PowerApps-TestEngine-main\src\PowerAppsTestEngine

D:\PowerApps-TestEngine-main\src\PowerAppsTestEngine> dotnet build
MSBuild version 17.4.0+18d5aef85 for .NET
  Determining projects to restore...
  Restored D:\PowerApps-TestEngine-main\src\PowerAppsTestEngine\PowerAppsTestEngine.csproj (in 272 ms).
  Restored D:\PowerApps-TestEngine-main\src\Microsoft.PowerApps.TestEngine\Microsoft.PowerApps.TestEngine.csproj (in 27
  2 ms).
  Microsoft.PowerApps.TestEngine -> D:\PowerApps-TestEngine-main\src\Microsoft.PowerApps.TestEngine\bin\Debug\net6.0\Mi
  crosoft.PowerApps.TestEngine.dll
  PowerAppsTestEngine -> D:\PowerApps-TestEngine-main\src\PowerAppsTestEngine\bin\Debug\net6.0\PowerAppsTestEngine.dll

Build succeeded.
    0 Warning(s)
    0 Error(s)

Time Elapsed 00:00:51.57
```

Figure 6.13 – Building the project

- In the generated bin\debug\PowerApps-TestEngine folder, you should be able to find the PowerShell script required to install the Playwright browsers: playwright.ps1. Please note that the destination folder of this script may have changed since the writing of this book. Therefore, refer to the official documentation or simply search for the playwright.ps1 file:

This PC › Windows (C:) › te3 › PowerApps-TestEngine › bin › Debug › PowerAppsTestEngine

Name	Date modified	Type	Size
Microsoft.Extensions.Logging.Abstractions.dll	10/23/2021 1:51 AM	Application extension	61 KB
Microsoft.Extensions.Logging.Configuration.dll	10/23/2021 1:50 AM	Application extension	27 KB
Microsoft.Extensions.Logging.Console.dll	10/23/2021 1:53 AM	Application extension	50 KB
Microsoft.Extensions.Logging.Debug.dll	10/23/2021 1:50 AM	Application extension	18 KB
Microsoft.Extensions.Logging.dll	10/23/2021 1:50 AM	Application extension	44 KB
Microsoft.Extensions.Logging.EventLog.dll	10/23/2021 1:50 AM	Application extension	24 KB
Microsoft.Extensions.Logging.EventSource.dll	10/23/2021 1:53 AM	Application extension	33 KB
Microsoft.Extensions.Options.ConfigurationExte...	10/23/2021 1:50 AM	Application extension	23 KB
Microsoft.Extensions.Options.dll	10/23/2021 1:50 AM	Application extension	58 KB
Microsoft.Extensions.Primitives.dll	10/23/2021 1:51 AM	Application extension	40 KB
Microsoft.Playwright.dll	6/13/2023 9:50 AM	Application extension	920 KB
Microsoft.PowerApps.TestEngine.dll	7/25/2023 10:19 PM	Application extension	158 KB
Microsoft.PowerApps.TestEngine.pdb	7/25/2023 10:19 PM	VisualStudio.pdb.dee...	504 KB
Microsoft.PowerFx.Core.dll	6/1/2023 3:22 PM	Application extension	979 KB
Microsoft.PowerFx.Interpreter.dll	6/1/2023 3:22 PM	Application extension	289 KB
Microsoft.PowerFx.Transport.Attributes.dll	6/1/2023 3:22 PM	Application extension	20 KB
Newtonsoft.Json.dll	11/24/2022 3:10 AM	Application extension	696 KB
playwright.ps1	6/13/2023 9:41 AM	Windows PowerShell ...	1 KB
PowerAppsTestEngine.deps.json	8/11/2023 1:33 PM	Archivo de origen JS...	39 KB
PowerAppsTestEngine.dll	8/11/2023 1:33 PM	Application extension	14 KB
PowerAppsTestEngine.exe	8/11/2023 1:33 PM	Application	145 KB

Figure 6.14 – Playwright PowerShell script

- To do so, execute the following command:

```
.\bin\Debug\PowerAppsTestEngine\playwright.ps1 install
```

Here is a screenshot for reference:

Figure 6.15 – Executing Playwright PowerShell install script

And that's it! Although we have not yet configured our config file to launch tests, to verify that the binary is working, we can attempt to run it from the same folder where we installed the Playwright PowerShell script. If everything is in order, you should see a message stating Couldn't find a project to run. This means we need to create our configuration file:

```
PS C:\te3\PowerApps-TestEngine\bin\Debug\PowerAppsTestEngine> dotnet run
Couldn't find a project to run. Ensure a project exists in C:\te3\PowerApps-TestEngine\bin\Debug\PowerAppsTestEngine, or
 pass the path to the project using --project.
PS C:\te3\PowerApps-TestEngine\bin\Debug\PowerAppsTestEngine>
```

Figure 6.16 – Executing Test Engine without a config file

We now have all the necessary elements in place to begin experimenting with Test Engine, which we will do in the next chapter, revisiting the sample that we used for Test Studio.

Summary

In this chapter, we have provided an overview of the new Test Engine project related to testing automation in Power Apps. As the *project is still in its experimental phase*, we have discussed how it addresses some of the limitations of Test Studio by maintaining Power Fx syntax and compatibility with Test Studio, allowing for *reusing tests recorded in Test Studio.*

We also discussed the additional features offered by Test Engine, such as connector mocking, screenshots, and video recording, as well as the limitations of Test Studio that Test Engine covers, such as component testing and nested galleries.

Based on Playwright, Test Engine has some limitations, such as a less user-friendly interface (command line) compared to a visual low-code tool and currently not supporting model-driven apps. However, the development team has plans to continue improving and expanding the capabilities of Test Engine, including integration with CI/CD processes.

Additionally, Test Engine is not yet integrated with Power Apps Studio and requires certain prerequisites to be ready to start working with it, as covered in this chapter. In the next chapter, we will cover how to perform tests with Test Engine.

Working with Test Engine

In the previous chapter, we provided an overview of Test Engine, discussing its capabilities, benefits, and current limitations. We also covered the prerequisites and build process required to create the Test Engine executable on your local machine.

Now that Test Engine is set up on your system, in this chapter, we will dive into the details of utilizing it for testing your Power Apps. We will cover how to work with the sample apps and test plans provided in the TestEngine GitHub repository (`https://github.com/microsoft/PowerApps-TestEngine`) so you can get first-hand experience using the tool.

The sample solutions provide a great way to get up and running quickly. You can import them into your Power Apps environment and execute the corresponding test plan YAML files against them. The samples demonstrate key functionalities such as taking screenshots, creating robust test plans in YAML, and more.

In addition to working with the provided samples, we will also discuss how to create your own new test plans tailored to your custom apps. Authoring tests in YAML will be covered, along with best practices for structuring your test plans. You will learn how to reference controls, configure test settings, parameterize your tests, and more.

Finally, we will cover how tests recorded in Power Apps Test Studio can be exported and executed directly in Test Engine without any modifications required. This provides a bridge for reusing your existing recorded test assets as you adopt Test Engine for automated testing needs.

By the end of this chapter, you will be well versed in the following:

- **Importing a sample solution**: In this section, we will explore how to import solutions provided in the Microsoft TestEngine GitHub repository. This will enable us to utilize these solutions as the destination for our tests.

- **Running tests and access results**: We will outline the process of executing the tests and delve into the interpretation of the results stored in the output folder.

- **Creating test plans with YAML**: We will provide a guide on how to create your own tests using YAML syntax.

- **Downloading and reusing recorded tests from Test Studio**: We will look at how to maximize the possibilities of Test Studio by downloading the tests created therein and executing them with Test Engine.

- **Using Power Platform Command-Line Interface (PAC CLI to execute tests defined in a Test Plan file**: In the appendix, we have introduced a new functionality in preview that allows the use of PAC CLI to execute Power Apps test plans.

With these skills, you will be equipped to start leveraging Test Engine's capabilities for all your real-world Power Apps testing needs. The upcoming chapters will build on this foundation and cover additional tips, tricks, and best practices.

Technical requirements

Before jumping into the hands-on examples in this chapter, let's recap what should already be in place from the previous chapter:

- The Test Engine executable should be built on your local machine. *Chapter 6* covered installing prerequisites such as the **.NET Core SDK** and then compiling the Test Engine source code from GitHub.

> **Note**
>
> As mentioned in *Chapter 6*, Microsoft has recently announced the new release of the PAC client, which now includes a new `tests` command. This essentially runs an embedded Test Engine executable. Although it is still in preview at the time of writing this book, it presents an alternative way to execute the tests that we craft in this chapter. We will delve into this new method in an *Appendix* section at the end of this chapter.

- We will need to import the `BasicGallery_1_0_0_2.zip` sample solution, which can be found in the TestEngine repository. For convenience, it has also been added to the repository for this chapter. Additionally, we will utilize the test plan available in the TestEngine repository as a template, so ensure the folder where you downloaded the TestEngine repository is readily accessible.

- You will also have access to several YAML test plans and configurations, which can be used as starters in the GitHub repository dedicated to this chapter (`https://github.com/PacktPublishing/Automate-Testing-for-Power-Apps/tree/main/chapter-07`). Additionally, the solution containing the canvas app created in *Chapter 5* is available there.

Please ensure these are in place before proceeding with the exercises in this chapter. If not, revisit *Chapter 6* to get Test Engine set up on your system and import the samples.

With the samples ready in your environment and Test Engine configured, you are set to start authoring and running test plans as we work through the examples in this chapter. Let's dive in and see Test Engine in action!

Importing a sample solution

Let's start with our actual hands-on work with Test Engine. For that, we are going to use one of the samples provided in the TestEngine GitHub repository (`https://github.com/microsoft/PowerApps-TestEngine`).

For now, we are going to use the *basic gallery sample* located in the GitHub repository. You can find the solution to import in the `samples` folder, and inside there is a ZIP file named `basic_gallery.zip` that we need to import into our environment in order to follow the steps for the initial basic test.

To import the solution into our environment, we will go to `https://make.powerapps.com`, and then, from there, access the **Solutions** section from the left-hand navigation.

Once we are in the **Solutions** section, we can use the **Import solution** button to browse and import the ZIP file located in the `samples/basicgallery` folder. In *Figure 7.1*, we can check the imports screen in Power Apps Studio.

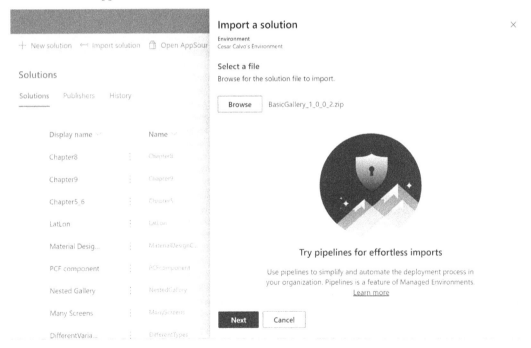

Figure 7.1 – Importing the BasicGallery solution

Once we have imported the solution, we will start creating the first configuration file to perform the tests.

For that, we will use the initial template (`config.json` as a configuration file base in `src/PowerAppsTestEngine`) in the folders that we downloaded from the TestEngine GitHub repository in *Chapter 6*. We will go to the `PowerAppsTestEngine` folder inside the `src` folder in the downloaded repository (`pathtodownloadedfolder/src/PowerAppsTestEngine`). If you have already compiled it as indicated in *Chapter 6*, you will be able to execute the tests from this folder. But let's start by creating our own configuration file to initiate the process.

We will take the `config.json` file as a starting point to create our own `config.dev.json` example (both files are provided in the chapter repository so you can choose to work directly with the `config.dev.json` file provided there) in the `PowerAppsTestEngine` folder.

Here, we will need to specify certain parameters such as the *ID of the environment* where we imported the application, and the *ID of the tenant*: both parameters can be found when logged in to the environment by clicking on the gear icon (**Settings**) in the top-right corner, and then clicking on **Session details**.

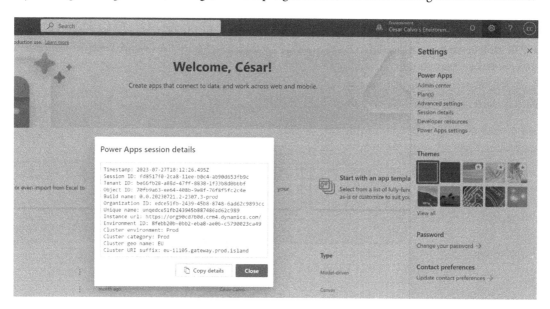

Figure 7.2 – Session details

We will need to indicate the file that actually contains the *test instructions*, and also the *output directory* that we will use to store our results, which might include screenshots and videos. So, we need to choose a disk and folder where we want to keep these results.

For instance, I am using the initial directory where I downloaded the GitHub project to point to the YAML file containing the test instructions, and I am using a different disk to store the test results, as you can see in the following figure:

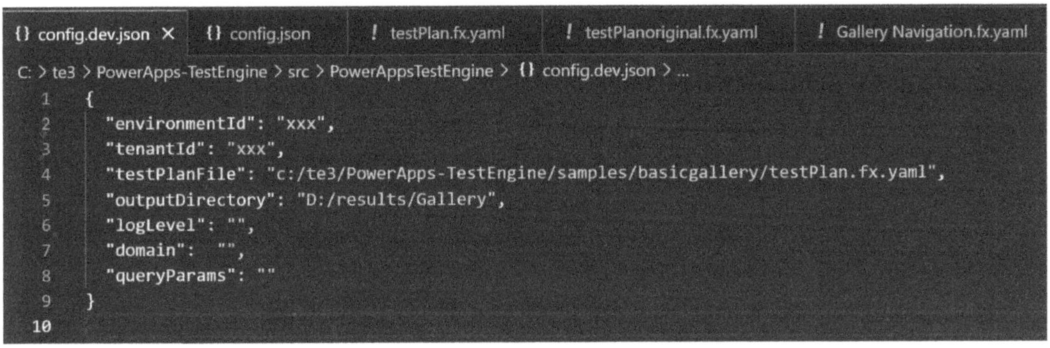

Figure 7.3 – Config file (config.dev.json)

We will save the configuration file as `config.dev.json`, since this is the naming convention that Test Engine expects for the configuration files used to run the tests specified within.

Before diving into the YAML file containing the test instructions, we first need to establish some environment variables. These will be employed to authenticate Test Engine when executing the tests. There are two ways to accomplish this task: you can either use Command Prompt or you can leverage the Windows **graphical user interface (GUI)**.

Using Command Prompt

First, let's use Command Prompt:

1. Open a new Command Prompt as Administrator.

2. Type the following command and press *Enter*:

    ```
    setx <variable_name> "<variable_value>"
    ```

3. Replace `<variable_name>` with the actual name of the variable (e.g., `user1Email`) and `<variable_value>` with the value to assign to it (e.g., `example@email.com`).

4. For example, this is how to create `user1Email` and `user1Password`:

    ```
    setx user1Email "example@email.com"
    setx user1Password "your_password_here"
    ```

> **Note**
>
> Please note that the new environment variables will only be available in future Command Prompt sessions.

This is how to check whether the `user1Email` and `user1Password` environment variables are set:

1. Open a new Command Prompt window.

2. Type the following command and press *Enter*:

   ```
   echo %<variable_name>%
   ```

   Replace `<variable_name>` with the actual name of the variable you want to check (e.g., `user1Email` or `user1Password`).

   For example, this is how to check `user1Email` and `user1Password`:

   ```
   echo %user1Email%
   echo %user1Password%
   ```

If the variables are set, their values will be displayed in Command Prompt.

Using the Windows GUI

Now, let's do the same using the Windows GUI:

1. Press *Win* + *I* to open **Windows Settings** and then navigate to **About | Device Specifications | Advanced System Settings**.

2. Click **Environment Variables** in the **Advanced** tab.

3. Under **User variables**, click **New** and enter the variable name and value (i.e., `user1Email`).

4. Repeat to create `user1Password`.

You will see a result similar to *Figure 7.4*, reflecting that the variables have been created. When you start a new Command Prompt session, you will then be able to use these variables.

Figure 7.4 – Environment variables for user credentials

Having confirmed our configuration for the tests and the necessary variables, let's delve into the syntax used in the test plans provided in the Test Engine repository.

Checking the YAML file

Let's start by checking the YAML file in the `samples` folder from the repository, and analyzing the sections and elements available there.

> **Note**
>
> In this section, we will utilize `testPlan.fx.yaml`, which is available in either the TestEngine repository (`samples/basicgallery`) or in the chapter repository of this book.

In the first section, we find the name and description of the **Test Suite** and the name of the user for the credentials that will log in to the app. We will specify the actual credentials for this persona in another section.

Very importantly, we need the *logical name of the app*. This is how Test Engine will identify the application to be tested. This is used when the Power Apps app is within a solution, as in this case.

If the application to be tested is not included in a solution, we would use `appID` instead, with the ID of the standalone app to be launched.

> **Note**
>
> In this case, the correct `appLogicalName` is already included in the sample. We will check later how to find it when using our own app.

In the following figure, we can see how `appLogicalName` and the other parameters are configured in our sample:

```
testSuite:
  testSuiteName: Basic Gallery
  testSuiteDescription: Verifies that you can interact with controls within a basic gallery
  persona: User1
  appLogicalName: new_galleryapp_80360
```

Figure 7.5 – YAML file: initial section

The subsequent section (`testCases`) is where the actual logic of the tests is outlined. As mentioned in the previous chapter, the language used for this logic is based on Power Fx, albeit with some current limitations, as we will elaborate on later.

From our sample, we can discern the following:

- We are instructing Test Engine to capture a screenshot at the beginning of the process
- We are confirming that the label, upon opening the app, matches the expected value
- We are navigating to the second and third rows and validating the related values

- We are altering the values of certain controls

- Finally, we are capturing another screenshot at the end of the process

Figure 7.6 shows what we find in our sample file in the `testSteps` section:

```
testCases:
  - testCaseName: Case1
    testSteps: |
      = Screenshot("basicgallery_loaded.png");
      Assert(Label1.Text = "Lorem ipsum 1", "Label should indicate first item in the gallery");
      Select(Label1);
      Assert(Index(Gallery1.AllItems, 2).Title1.Text = "Lorem ipsum 2", "Validate the label in the 2nd row of the gallery");
      Select(Index(Gallery1.AllItems, 2).NextArrow1);
      Assert(Label1.Text = "Lorem ipsum 2", "Label should be updated to indicate second item in the gallery");
      // Using the test studio syntax to select gallery item
      Select(Gallery1, 2);
      Select(Gallery1, 3, NextArrow1);
      Assert(Label1.Text = "Lorem ipsum 3", "Label should be updated to indicate third item in the gallery");
      // Using SetProperty to change the values on the controls
      SetProperty(Label1.Text, "End of the test");
      SetProperty(Index(Gallery1.AllItems, 2).Title1.Text, "End of the test");
      Assert(Index(Gallery1.AllItems, 2).Title1.Text = "End of the test", "Label in the gallery should be updated");
      Screenshot("basicgallery_end.png");
```

Figure 7.6 – YAML file: testSteps section

Let's now focus on the `testSettings` section:

- `locale` (used in the example): Determines the language and region settings for the test. In the example, it is set to `en-US` for English (United States).

- `recordVideo` (used in the example): Specifies whether to capture video of the test execution. In the example, it is set to `true`, meaning a video will be recorded.

- `headless` (used in the example): Indicates whether the test should run without a visible user interface (headless mode). In the example, it's set to `false` to be able to see the operations executed.

> **Note**
>
> In the sample YAML file, ensure that the `headless` parameter is set to `false`. This allows visibility into the operations being performed.

- `browserConfigurations` (in the example): Lists the browsers and configurations for running the tests. In this example, the test will run on the Chromium browser.

- `timeout`: Sets the maximum duration in milliseconds for a test step before timing out.

- `pollingInterval`: Defines the interval in milliseconds between polling attempts when waiting for a condition.

These parameters help configure the test suite's settings, allowing easier management and modification of the test execution without changing the actual test code.

```
testSettings:
  locale: "en-US"
  recordVideo: true
  headless: false
  browserConfigurations:
    - browser: Chromium
environmentVariables:
  users:
    - personaName: User1
      emailKey: user1Email
      passwordKey: user1Password
```

Figure 7.7 – YAML file: test settings and environment variables section

The `environmentVariables` section in the YAML file defines variables to be used during test execution. These variables can store values such as credentials that should not be hardcoded into the tests.

We have the option of configuring a `filePath` pointing to a separate YAML file containing all the environment variables. We have not done this in the example and are instead creating the needed authentication variables directly in the `users` section.

The `users` parameter is part of `environmentVariables` and contains user personas with these properties:

- `personaName`: A name identifying the user persona (for example, "User1").
- `emailKey`: A key mapping to the user's email address. This key accesses the actual email value from the environment variables configured earlier (described in the previous section).
- `passwordKey`: A key mapping to the user's password. This key accesses the actual password value from the configured environment variables.

Now that we have everything configured, it's time to start testing the **BasicGallery application** as intended from the beginning of this chapter.

Running tests and accessing results

To start testing, we need to open Command Prompt, change to the directory where the TestEngine GitHub repository was downloaded, and locate the `src` folder. Go into the `PowerAppsTestEngine` folder with the project that we want to run.

The key steps are as follows:

1. Open Command Prompt.
2. Navigate to `<downloaded_repo_path>\src\PowerAppsTestEngine`.

3.   Run the `PowerAppsTestEngine` project.

Figure 7.8 – Folder of the TestEngine project

This is the same folder in which we should have the `config.dev.json` file, as described in the initial requirements section. Note that in this file, we are specifying different folders to locate the test file and the output directory, in addition to the tenant ID and environment ID of the app.

Take note of the output directory you have specified, as we will inspect this folder after the test run.

Now, let's initiate the test execution. In our case, we have only one sample test case, but note that we could have multiple test cases from the test plan within the same file. Start the execution in Command Prompt using the following:

```
dotnet run
```

This will run the `PowerAppsTestEngine` project and execute the tests based on the YAML file, using the configuration defined in `config.dev.json`.

In *Figure 7.9*, we can observe how the execution of this command initiates a new session and launches the application with the test steps outlined in the corresponding section.

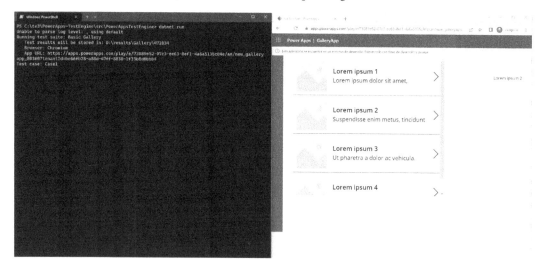

Figure 7.9 – Running Test Engine

Executing `dotnet run` will launch the Chromium browser (downloaded for *Playwright* in *Chapter 6*) and perform the test steps detailed in our YAML test case. Simultaneously, it will log the actions in Command Prompt, providing information such as the app URL, the utilized browser, the storage location for test results, and the outcomes from each test case. Command Prompt will display the progress and results of the automated test execution based on our configured test case.

Upon completion of the test execution, a summary will be presented in Command Prompt, displaying the total number of test cases run and the number that passed. It will also indicate the location of the TRX file (which will be discussed later) in the `output` folder specified in the configuration.

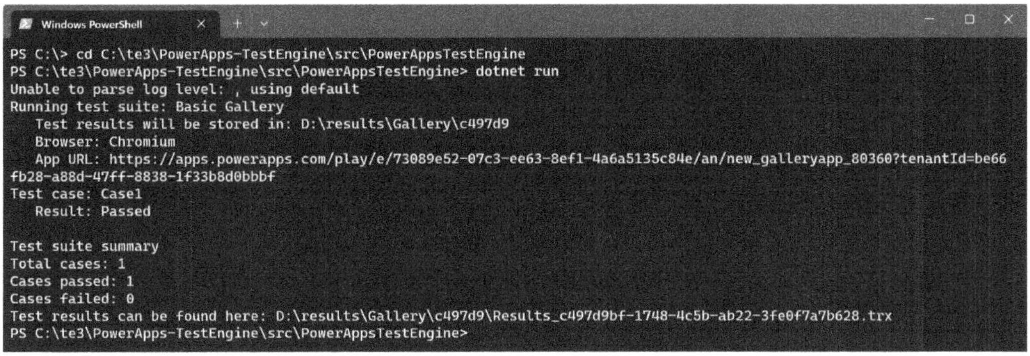

Figure 7.10 – Test Engine summary

In the upcoming section, we will analyze these outputs, which include the TRX file, videos, screenshots, and logs.

Inspecting test run results and recordings

The `output` folder is the location where we can find the generated videos, screenshots, logs, and other artifacts from the test run. Reviewing these output files will enable us to analyze the detailed results and understand exactly what transpired during the automated test execution. Let's start with the TRX file.

TRX file

As previously noted, one of the output files produced is the TRX file. This can be found in the output directory configured in our `config.dev.json` file. The test execution summary in Command Prompt indicates the specific location – in the example provided earlier, it is pointing to `D:\results\gallery\c497d9`. Please note, this is contingent upon the path configured in the `outputDirectory` parameter.

Inspecting this folder allows us to locate the TRX file generated from the test run. The TRX file contains detailed results data that can be consumed by other tools and integrated into CI/CD pipelines.

A **TRX** (**Test Results XML**) file is an XML-formatted file that stores the results of automated tests. These test results are typically generated when running automated tests using frameworks such as MSTest, NUnit, or xUnit in Visual Studio, or other testing tools integrated with Azure DevOps.

In the following figure, we can see the structure of the TRX file produced by this test execution.

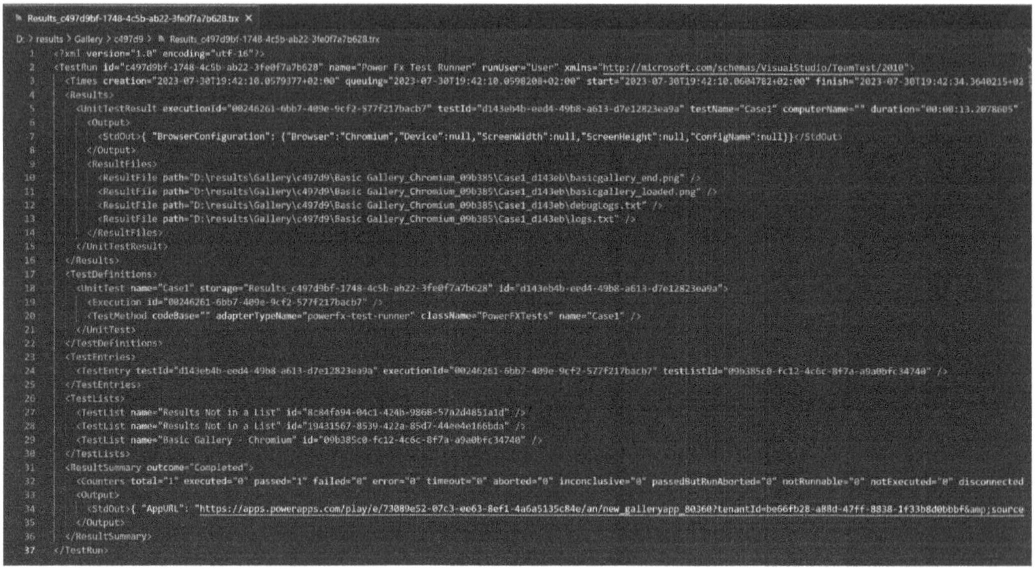

Figure 7.11 – XML structure of a TRX file

The TRX file serves as a standardized way to store test outcomes, including the following:

- **Test execution details**: Includes data such as test start time, end time, duration, and outcome (pass, fail, and inconclusive)

- **Test method information**: Contains data about individual test methods executed, such as name, namespace, and fully qualified name

- **Test run summary**: Provides an overview of the entire test run, including total tests executed, the number passed, failed, and inconclusive, and any error messages or stack traces from failures

- **Environment data**: Can capture information about the test environment such as machine name and user who started the test run

The generation of TRX files as output from test automation is crucial for seamless integration with Azure DevOps and **Continuous Integration/Continuous Deployment** (**CI/CD**) pipelines. TRX provides a standardized format for test results, enabling easier communication and information sharing across testing frameworks.

> **Note**
>
> At the time of writing this book, integration of Test Engine with **PAC CLI** (the command-line tool for managing and automating Power Apps development tasks) has been announced in the public preview phase. This will allow the use of TRX files as output in CI/CD pipelines leveraging PAC CLI commands to invoke Test Engine. We will explore how to take advantage of this tool later.

Another noteworthy output from our test execution is the videos and screenshots taken during the process. We'll examine these artifacts in the following section.

Videos and screenshots

To start examining them, recall that we configured the YAML file to record a video of the execution using the `recordVideo` parameter under `testSettings`:

```
recordVideo: true
```

This records the entire test execution and saves it as a `.webm` video in a subfolder under the root output folder, usually named after the browser and related to the test suite – in this case, `D:\results\Gallery\c497d9\Basic Gallery_Chromium_09b385`.

Name	Date modified	Type	Size
Case1_d143eb	7/30/2023 7:42 PM	File folder	
7b6454734b848349330cfe78d3cb4740.webm	7/30/2023 7:42 PM	WEBM File	419 KB
debugLogs.txt	7/30/2023 7:42 PM	Text Document	10 KB
logs.txt	7/30/2023 7:42 PM	Text Document	3 KB

> This PC › PROGS (D:) › results › Gallery › c497d9 › Basic Gallery_Chromium_09b385

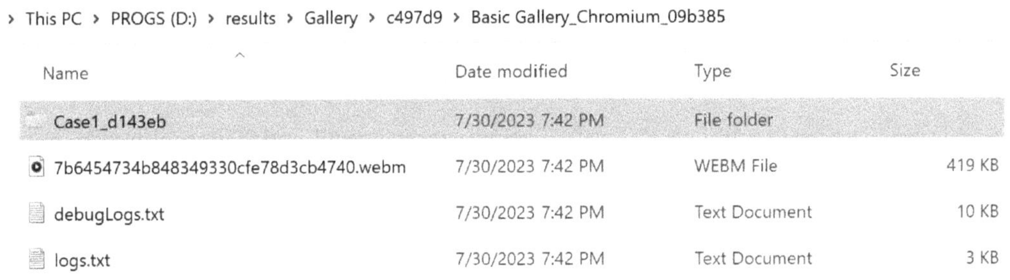

Figure 7.12 – Path of the recorded video

Opening this video in a media player allows for observing all test steps taken.

Additionally, in the `testSteps` section of our YAML file, we added `Screenshot` commands at the beginning and end:

```
Screenshot("basicgallery_loaded.png");
Screenshot("basicgallery_end.png");
```

These save screenshots to another subfolder related to the test case under the one mentioned earlier, named per the test case. In our example, it could be something like `D:\results\Gallery\c497d9\Basic Gallery_Chromium_09b385\Case1_d143eb`, as you can see in the following figure:

This PC › PROGS (D:) › results › Gallery › c497d9 › Basic Gallery_Chromium_09b385 › Case1_d143eb

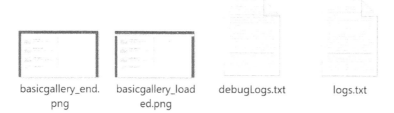

basicgallery_end.   basicgallery_load   debugLogs.txt   logs.txt
png   ed.png

Figure 7.13 – Screenshots generated during Test Engine execution

> **Note**
>
> It's worth noting that the screenshot paths are referenced in the TRX file. So, when integrating with Azure DevOps, these screenshots can accompany the test reports, providing helpful visual context for the test results.

Another significant outcome from our tests, which can be examined for more detailed information depending on the level of granularity, is the logs. We will address this in the upcoming section.

Logs

Test Engine generates log files at multiple levels of granularity and detail, and they are also referenced in the TRX file for the `TestCase` level:

- Root folder logs cover test configuration details such as output folder, locale, and app domain
- Test suite folder logs capture the execution of the full test suite and final summary
- Test case folder logs contain specifics about each test case execution

Within these folders, there are **logs** and **debugLogs** files providing different levels of detail:

- `logs` files are more concise, focusing on critical information such as browser configuration, test setup, target URL navigation, and test suite summary
- `debugLogs` files include more granular actions during test case execution such as screenshots, assertions, selections, property settings, and potential code exceptions

Inspect *Figure 7.14* to view a `debugLogs` file at the `TestCase` level.

```
------------------------------------------------------------------------
RUNNING TEST CASE: Case1
------------------------------------------------------------------------

Attempting:

{
 Screenshot("basicgallery_loaded.png");
  Assert(Label1.Text = "Lorem ipsum 1", "Label should indicate first item in the gallery");
  Select(Label1);
  Assert(Index(Gallery1.AllItems, 2).Title1.Text = "Lorem ipsum 2", "Validate the label in the 2nd row of the gallery");
  Select(Index(Gallery1.AllItems, 2).NextArrow1);
  Assert(Label1.Text = "Lorem ipsum 2", "Label should be updated to indicate second item in the gallery");
  // Using the test studio syntax to select gallery item
  Select(Gallery1, 2);
  Select(Gallery1, 3, NextArrow1);
  Assert(Label1.Text = "Lorem ipsum 3", "Label should be updated to indicate third item in the gallery");
  // Using SetProperty to change the values on the controls
  SetProperty(Label1.Text, "End of the test");
  SetProperty(Index(Gallery1.AllItems, 2).Title1.Text, "End of the test");
  Assert(Index(Gallery1.AllItems, 2).Title1.Text = "End of the test", "Label in the gallery should be updated");
  Screenshot("basicgallery_end.png");
}

------------------------------

Executing Screenshot function.

Successfully finished executing Screenshot function.

Run Javascript:
PowerAppsTestEngine.getPropertyValue({"controlName":"Label1","index":null,"parentControl":null,"propertyName":"Text"}).then((propertyValue) =>
JSON.stringify(propertyValue))

------------------------------
```

Figure 7.14 – A debugLogs file at the TestCase level

After experimenting with the sample provided by the TestEngine repository, we can now edit another one to integrate our own logic and connect it with our sample contacts application. This will allow us to customize the sample and link it to an existing project, gaining hands-on experience modifying these resources. By leveraging the Test Engine samples as a starting point, we can rapidly prototype and iterate as we build out our application's test automation framework.

Creating test plans with YAML

Assuming you have followed the guidelines outlined in *Chapter 5*, you should now have the **Gallery with Contacts** application that we created – alternatively, you can find the solution with the application and the data to import in the GitHub repository for this book (`https://github.com/PacktPublishing/Automate-Testing-for-Power-Apps/tree/main/chapter-07`). This application features a gallery that draws its data from the `contacts` table and is designed to resemble *Figure 7.15*.

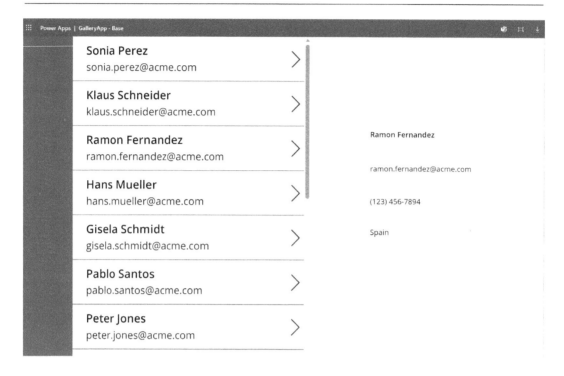

Figure 7.15 – Sample application with Gallery control and contacts table

Now, we will construct a new YAML file that outlines the necessary steps to do the following:

1. Generate an initial screenshot displaying the first element.

2. Navigate to the fifth element, verify that the associated email address matches our expectations (in our case, it's `gisela.schmidt@acme.com`, but bear in mind that this might differ if you have not imported the contacts data or if you had previously imported different data), and capture a new screenshot.

3. Navigate to the third element, ensure the full name displayed in the details pane matches our expectations (in my case, `Ramon Fernandez`), and capture another screenshot.

Before proceeding, let's review our `config.dev.json` file to ensure the correct environment, tenant, and location of the new YAML file containing the test case are specified. See *Figure 7.16* for the new configuration.

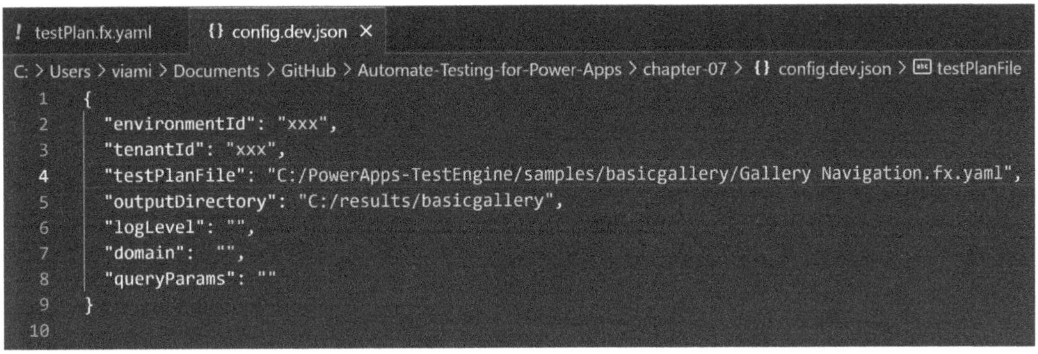

```
!  testPlan.fx.yaml       {} config.dev.json  ×

C: > Users > viami > Documents > GitHub > Automate-Testing-for-Power-Apps > chapter-07 > {} config.dev.json > ▦ testPlanFile
  1    {
  2        "environmentId": "xxx",
  3        "tenantId": "xxx",
  4        "testPlanFile": "C:/PowerApps-TestEngine/samples/basicgallery/Gallery Navigation.fx.yaml",
  5        "outputDirectory": "C:/results/basicgallery",
  6        "logLevel": "",
  7        "domain":  "",
  8        "queryParams": ""
  9    }
 10
```

Figure 7.16 – Configuration in config.dev.json

As you're now pointing to a new YAML file (here, `Gallery Navigation.fx.yaml`, also available in the repository for our current chapter), consider specifying a distinct results folder to differentiate from previous samples. Also, ensure that `tenantid` and `environmentid` are configured correctly. Remember, you can find both values by opening the **Settings | Session details** dialog within the environment.

Next, we'll edit the YAML file to create the necessary steps for the desired functionality. Let's start with the `testSuite` section, ensuring the correct `appId` or `appLogicalName` is specified. In the following figure, we have highlighted the `appLogicalName` to test:

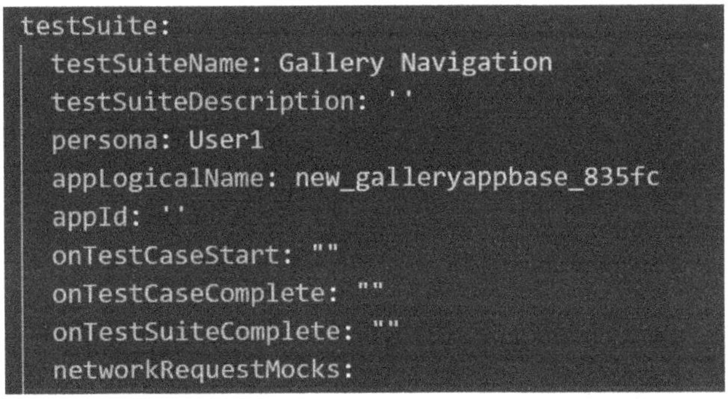

```
testSuite:
  testSuiteName: Gallery Navigation
  testSuiteDescription: ''
  persona: User1
  appLogicalName: new_galleryappbase_835fc
  appId: ''
  onTestCaseStart: ""
  onTestCaseComplete: ""
  onTestSuiteComplete: ""
  networkRequestMocks:
```

Figure 7.17 – Test suite parameters in YAML file

Given the requirement for either `appLogicalName` or `appId` of our application, let's explore how we can obtain both.

To get the app ID of a canvas Power Apps app, follow these steps:

1. **Select the environment**: Ensure you select the environment where your application resides.

2. **Find your app**: Click on **Apps** near the left edge to see a list of available apps in your environment.

3. **Go to the app details**: Locate the canvas Power Apps app you want to get the app ID for, click on the **Details** button or the three dots next to the app, and select **Details** from the **Context** menu.

4. **Get the app ID**: On the app **Details** page, you'll find the app ID at the bottom.

To get the logical name of a canvas Power Apps app included in a solution, follow these steps:

1. **Access the solution**: Click on **Solutions** in the left navigation pane, and then open the solution where your canvas Power Apps app is available.

2. **Locate your app**: Find the canvas Power Apps app you want to get the logical name for within the solution.

3. **Get the logical name**: The logical name should be displayed next to the app in the solution. If it's not visible, you can try clicking on the app to see whether the name is displayed in the app details or properties.

> **Note**
>
> The logical name might not be the same as the display name of the app. It is a more technical identifier used for referencing the app in certain scenarios, while the display name is a more human-readable name for the app.

Once we've completed the `testSuite` section, we'll inspect the `testSettings` section and implement similar configurations to those used in our previous test. This will include recording the video and utilizing the environment settings for authentication, as you can see in *Figure 7.18*.

```
testSettings:
  filePath:
  browserConfigurations:
  - browser: Chromium
  recordVideo: true
  headless: false
  enablePowerFxOverlay: false
  timeout: 30000
  workerCount: 10
environmentVariables:
  filePath:
  users:
  - personaName: User1
    emailKey: user1Email
    passwordKey: user1Password
```

Figure 7.18 – Test settings in the YAML file

Excellent, let us now proceed to create the necessary steps in the `testCase` section. The following code will serve as our foundation, and we will subsequently review the meaning of each line. However, please feel free to experiment with Power Fx syntax and validate new elements of the interface:

```
testCases:
  - testCaseName: Check Row5 and 3 (Email and Full Name)
    testCaseDescription: ''
    testSteps: |
      =
      Screenshot("gallerycontacts_loaded.png");
      Select(Gallery1, 5, NextArrow1);
      Assert(Index(Gallery1.AllItems, 5).Subtitle1.Text = "gisela.
      schmidt@acme.com", "Email in row 5 does not correspomd");
      Screenshot("gallerycontacts_contacts5.png");
      Select(Gallery1, 3, NextArrow1);
      Assert(Index(Gallery1.AllItems, 3).Title1.Text = "Ramon
      Fernandez", "Full name in row 3 does not correspond");
      Screenshot("gallerycontacts_contacts3.png");
```

After defining the test case name based on the navigation to specific rows within the gallery for value checking, it is recommended to add a test description.

Then, initiate the actual test steps, as follows:

1. First, take a screenshot of the initial screen.

2. Then, navigate to the fifth row using the `Select` operation.

3. Once on the fifth element, use an `Assert` instruction to verify that the displayed email corresponds to the expected value.

4. Take a new screenshot of the application to validate this step.

5. Next, navigate to the third row and verify that the full name corresponds to the expected value using the `Assert` instruction.

6. Finally, take another screenshot to visually verify the value.

When executing the test with `dotnet run`, we should expect to see a result similar to *Figure 7.19*, indicating that the test case has passed successfully. The result will include a summary of the operations performed:

```
PS C:\te3\PowerApps-TestEngine\src\PowerAppsTestEngine> dotnet run
Unable to parse log level: , using default
Running test suite: Gallery Navigation
    Test results will be stored in: D:\results\GalleryContacts\324966
    Browser: Chromium
    App URL: https://apps.powerapps.com/play/e/8febb20b-0bb2-eba8-ae06-c5790023ca49/an/new_galleryappbase_835fc?tenantId=
be66fb28-a88d-47ff-8838-1f33b8d0bbbf
Test case: Check Row5 and 3 (Email and Full Name)
    Result: Passed

Test suite summary
Total cases: 1
Cases passed: 1
Cases failed: 0
Test results can be found here: D:\results\GalleryContacts\324966\Results_324966b7-b41f-4939-bdda-3d1a0d06af8d.trx
PS C:\te3\PowerApps-TestEngine\src\PowerAppsTestEngine>
```

Figure 7.19 – Summary output of test execution

Next, we will check the folders containing the images and video produced by the test. As previously established, we can find the output in the folder specified in the `config.dev.json` file. In this case, the folder can be found at `D:\results\GalleryContacts`.

Figure 7.20 – Folders structure in the Test Engine output

We can locate the folder mentioned in the output summary, which, in our case, is the `324966` folder (remember that you will get a different number).

Inside this folder, we will find the same structure as that of the initial test case. This includes the following:

- The root folder with the logs related to the test suite and the TRX file that we previously discussed

- A subfolder with the logs of the test suite and the video, as this will include all the test cases included in the test suite

- Another subfolder for each of the test cases, containing the screenshots and logs related to the test case level

One interesting option of Test Engine is the integration with Test Studio. We can export test cases created in Test Studio and adapt them to run with Test Engine by modifying certain things. Let's explore this option in the next section.

Downloading and reusing recorded tests from Test Studio

First, we'll describe how to generally download a YAML file from Test Studio and adapt it for use in Test Engine. Then, we'll examine how to download and adapt the specific sample we created in Test Studio in *Chapter 5*.

To download a YAML test file from Test Studio for use in Test Engine, follow these steps:

1. In Test Studio, download recorded tests using the **Download suite** button. Choose the desired test suite if you have multiple. Alternatively, use the **Download** button under each test suite.

2. In the Test Engine `config.dev.json` file, update the `testPlanFile` parameter to point to the downloaded YAML file (for example, `../../samples/basicgallery/Gallery Navigation.fx.yaml`). Also, update the user credentials in the `environmentVariables` section if using a different tenant or environment.

3. Run the test in Test Engine using the `dotnet run` command.

By following these general steps, you can download any YAML file from Test Studio and adapt it for testing Power Apps in Test Engine.

Next, we'll look specifically at downloading and adapting the sample we created in *Chapter 5*.

We will navigate to the **GalleryApp** application (or any other name chosen) and edit it. Then we can click into the **Advanced tools** section and click on **Open Tests**. This way, we will get the Test Studio screen that we visited in *Chapter 5*.

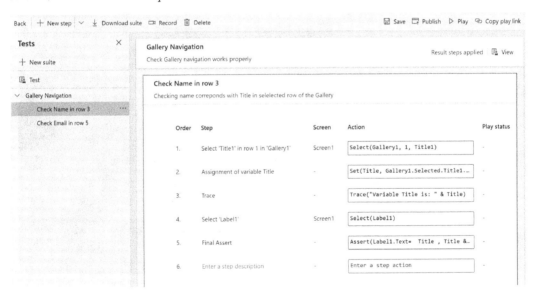

Figure 7.21 – Downloading the test suite YAML file from Test Studio

From this point, you can click the **Download suite** button to download the YAML file containing the test cases you've created. A good practice might be renaming this file to easily identify it as downloaded from Test Studio. For instance, you could name it `GalleryAppNavigation.TestStudio.yaml`.

The initial structure of the YAML file downloaded for this sample can be seen in the following figure:

```yaml
! Gallery Navigation.TestStudio.yaml  ✕

C: > Users > viami > Desktop > ! Gallery Navigation.TestStudio.yaml
 1   testSuite:
 2     testSuiteName: Gallery Navigation
 3     testSuiteDescription: ''
 4     persona: User1
 5     appLogicalName: new_galleryappbase_835fc
 6     appId: ''
 7     onTestCaseStart: ""
 8     onTestCaseComplete: ""
 9     onTestSuiteComplete: ""
10     networkRequestMocks:
11     testCases:
12     - testCaseName: Check Row3 (Name)
13       testCaseDescription: ''
14       testSteps: |
15         =
16         Select(Gallery1, 3, NextArrow1);
17         Set(Title, Gallery1.Selected.Title1.Text);
18         Trace("Title1 : " & Title);
19         Select('Full Name');
20         Assert('Full Name'.Text=  Title , Title & " does not correspond with  " & 'Full Name'.Text);
21     - testCaseName: Check Row5 (Email)
22       testCaseDescription: ''
23       testSteps: |
24         =
25         Select(Gallery1, 5, NextArrow1);
26         Set(Subtitle, Gallery1.Selected.Subtitle1.Text);
27         Trace("Gallery Subtitle : " & Subtitle);
28         Select(Email);
29         Assert(Email.Text=  Subtitle , Subtitle & " does not correspond with  " & Email.Text);
30   testSettings:
31     filePath:
32     browserConfigurations:
33     - browser: Chromium
34       device:
35       screenWidth: 0
36       screenHeight: 0
37     locale: en-US
38     recordVideo: true
39     headless: true
40     enablePowerFxOverlay: false
41     timeout: 30000
42   environmentVariables:
43     filePath:
44     users:
45     - personaName: User1
46       emailKey: user1Email
47       passwordKey: user1Password
```

Figure 7.22 – YAML file downloaded from Test Studio

We will now be able to modify this file with our own Power Fx instructions or adjust several parameters as needed.

Please keep in mind that this project is still in the experimental phase. Consequently, it's possible that the YAML instructions downloaded may not completely align with the Power Fx syntax recognized by Test Engine.

As a result, you might find it necessary to fine-tune some instructions. We hope you enjoy testing and exploring this feature!

Appendix – using PAC CLI to execute tests defined in a Test Plan file

In this appendix, we will explore the use of **PAC CLI**, also known as **pac cli**, to execute tests defined in a Test Plan file for canvas Power Apps apps. We've been working with the Test Engine executable that we compiled in *Chapter 6*, but a recent announcement introduces a new way to execute these tests using the `pac tests run` command included in the pac client.

> **Note**
>
> This feature has just been announced and is currently in the public preview phase, which means it's subject to change. Please refer to the official documentation for any updates (`https://learn.microsoft.com/en-us/power-platform/developer/cli/reference/tests`).

Before we begin, it's necessary to install the *PAC CLI*. You can use either or both of the following methods for installation:

- **Install using Power Platform Tools for Visual Studio Code**: This method enables the use of commands within a PowerShell terminal inside Visual Studio Code

- **Install Power Platform CLI for Windows**: This method enables the use of commands using Command Prompt or PowerShell on Windows 10 and Windows 11

For more details on installing the Power Platform CLI, refer to the official documentation (`https://learn.microsoft.com/en-us/power-platform/developer/cli/introduction`).

After installing *PAC CLI*, we need to make the `pac.exe` client accessible from anywhere in the command line. For that (assuming we are working on Windows), we need to cover the following steps.

Add the directory containing the pac.exe file to the system's PATH environment variable

To make the `pac.exe` client accessible from anywhere in the command line on Windows, add the directory containing the `pac.exe` file to the system's PATH environment variable. Follow these steps:

1. Locate the directory containing the `pac.exe` file. For example, it might be in `C:\PowerPlatformCLI\`.

2. Right-click on **This PC** or **My Computer** and select **Properties**.

3. Click on **Advanced system settings** on the left side of the **System Properties** window.

4. In the **System Properties** window, click on the **Environment Variables** button.

5. In the **Environment Variables** window, locate the **Path** variable in the **System variables** section and click on **Edit**.

6. In the **Edit environment variable** window, click on **New** and add the directory path containing the `pac.exe` file (e.g., `C:\PowerPlatformCLI\`).

7. Click **OK** to close the **Edit environment variable** window, then click **OK** again to close the **Environment Variables** window, and finally, click **OK** to close the **System Properties** window.

You should now be able to run the `pac.exe` client from any location in the command line. If the changes don't take effect immediately, you may need to restart your Command Prompt or PowerShell terminal.

To verify that we have the `pac tests run` command available, we need to check the version of PAC CLI we're using. You can do this by simply typing the `pac` command in Command Prompt. As of the time of writing this book, the latest version is `1.26.5`, and it includes this functionality in preview.

The output should list the available commands, including the `tests` command. If it doesn't, you may need to update your PAC CLI.

Assuming that we have our PAC CLI installed and accessible from everywhere in the command line, we will now cover how to run tests with PAC CLI.

Running automated tests for a Power Apps app with PAC CLI

Navigate to the directory containing your Test Plan file (note that, in this case, we won't need a `config.dev.json` file as we are passing the parameters normally written in this file as command parameters). We assume that you have set up your environment variables for `user1Email` and `user1Password` as described in the previous section.

If you want to store the output results in `D:\results\paccli`, an example command would be as follows:

```
pac tests run --test-plan-file testPlan.fx.yaml --environment-id
<your-environment-id> --tenant <your-tenant-id> --output-directory D:\
results\paccli
```

Replace `<your-environment-id>` with your app's environment ID and `<your-tenant-id>` with your app's tenant ID. These, along with the output directory, were managed by Test Engine in the `config.dev.json` file.

Executing the command will produce an output similar to what is shown in *Figure 7.23*, and the results we discussed in previous sections will be created in the chosen folder.

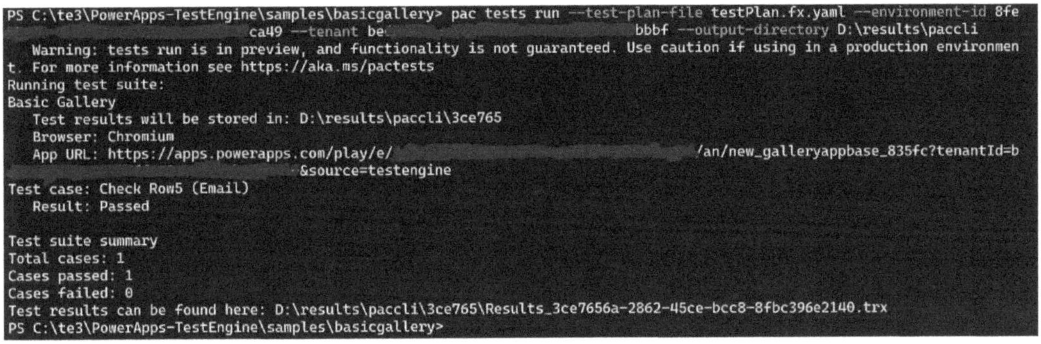

```
PS C:\te3\PowerApps-TestEngine\samples\basicgallery> pac tests run --test-plan-file testPlan.fx.yaml --environment-id 8fe
              ca49 --tenant be                                    bbbf --output-directory D:\results\paccli
   Warning: tests run is in preview, and functionality is not guaranteed. Use caution if using in a production environmen
t. For more information see https://aka.ms/pactests
Running test suite:
Basic Gallery
   Test results will be stored in: D:\results\paccli\3ce765
   Browser: Chromium
   App URL: https://apps.powerapps.com/play/e/                                    /an/new_galleryappbase_835fc?tenantId=b
              &source=testengine
Test case: Check Row5 (Email)
   Result: Passed

Test suite summary
Total cases: 1
Cases passed: 1
Cases failed: 0
Test results can be found here: D:\results\paccli\3ce765\Results_3ce7656a-2862-45ce-bcc8-8fbc396e2140.trx
PS C:\te3\PowerApps-TestEngine\samples\basicgallery>
```

Figure 7.23 – Executing tests with pac cli

As illustrated, this method can serve as an alternative means of executing automated tests for Power Apps. Although it lacks the extensibility of the Test Engine project – given that we have access to its source code – it is included in PAC CLI, which houses additional utilities for Power Platform projects and is likely an indispensable toolset for working with Power Platform.

Moreover, this approach opens new possibilities for automated testing within **Application Lifecycle Management** (**ALM**) processes, which we will explore further in *Chapter 10* of this book.

Summary

In this chapter, we dove deep into the practical usage of Test Engine. We started with a recap of the prerequisites and build process required to create the Test Engine executable on your local machine, as covered in the previous chapter. Then, we walked through a detailed hands-on guide on how to use Test Engine to test Power Apps based on sample apps and test plans from its GitHub repository.

The chapter covered how to import sample solutions into your Power Apps environment, execute test plans, inspect test run results and recordings, author new test plans in YAML, and reuse existing Test Studio assets. We also discussed the technical requirements and steps to run the first test with Test Engine using a sample app from the GitHub repository.

We provided a comprehensive explanation of how to import the solution and the initial requirements, create a configuration file, set environment variables, and understand the YAML file structure. We covered how to run the test, review the output files, understand the TRX file, and check the videos, screenshots, and logs generated by Test Engine.

The chapter also highlighted how to test with your own YAML file, download a YAML file from Test Studio, and adapt it for use in Test Engine. We also mentioned the newly announced integration of Test Engine with pac cli, which allows the use of this tool to execute automated test plans for Power Apps.

By this point, you should be prepared to leverage the capabilities of Test Engine for real-world Power Apps testing needs. In the following chapters, we will address some key limitations of Test Studio with Test Engine. But before we do, let's explore, in *Chapter 8*, an alternative method for automating the testing of Power Apps: **Power Automate Desktop** (**PAD**). We will start with the basic concepts of PAD to demonstrate how it can be utilized for automating tests of Power Apps.

8

Testing Power Apps with Power Automate Desktop

In the preceding chapters, we delved into two methods of testing Power Apps: Test Studio and Test Engine, with the latter having recently been integrated into the PAC CLI. Now, we'll turn our attention to an alternative testing method that could be well suited for specific scenarios: **Robotic Process Automation (RPA)**. RPA is often regarded as a key component in digital transformation strategies, presenting itself as a viable alternative in testing certain integrated scenarios. Such scenarios could include instances where it is necessary to transfer values or verify final values across different interfaces, such as websites or legacy applications.

In this scenario, **Power Automate Desktop (PAD)** enters the stage. As an important tool of the Power Automate suite, PAD provides a robust environment to create RPA workflows or flows that simulate user interactions with an application. With the ability to perform a multitude of actions, PAD can automate complex actions that would be time-consuming and error-prone if performed manually.

In the context of Power Apps testing, this book delves into PAD as an alternative methodology. Our focus remains on handling components that, though external to Power Apps itself, are nonetheless integral to the overall solution. While this method is not likely to be the preferred approach – especially considering the availability of tools such as Test Studio and Test Engine, which we discussed previously – it does offer an alternate perspective. This can be particularly beneficial when there is an interaction with other components of the Power Platform ecosystem, such as model-driven apps, Power Pages, or Power Virtual Agents.

> **Note**
>
> It is important to remark that while PAD can prove to be a tool for testing in certain circumstances, it is not normally the optimal choice. It represents just one of many testing approaches available, and its effectiveness can greatly vary depending on the specifics of the use case. Therefore, evaluating the requirements and constraints of your particular scenario is crucial before deciding on the most suitable testing method.

To exemplify this alternative, throughout this chapter, we'll be working with two different examples: a **model-driven app** and a **canvas app**. It's important to recall that model-driven apps are currently not supported by Test Studio or Test Engine. Although there are announcements that they will be covered by Test Engine in the future, this feature has not been implemented yet. Hence, **PAD**, which replicates user interactions, presents us with an opportunity to address this scenario.

We will explore the core concepts of PAD and its application in testing. We will discuss how RPA tools such as PAD can serve as an alternative in automating UI testing,

The highlight of this chapter will be a practical walk-through on how to create a desktop flow that enables UI testing for a simple model-driven app. These same techniques can be applied to canvas apps or Power Pages.

This journey will comprise the following sections:

- **An introduction to Power Automate Desktop and its significance in business process testing**: This section provides an overview of RPA and PAD, elaborating on how they can serve as alternative methods for automated testing

- **Key considerations while using PAD as a testing tool**: In this section, we'll discuss the vital aspects to consider when contemplating the use of PAD for testing

- **Benefits of using PAD for end-to-end business process testing**: In this section, we'll explore the scenarios where PAD can be most effectively applied and the associated benefits

- **A step-by-step guide to creating a desktop flow for testing a model-driven app and a canvas app**: In this practical segment, we'll demonstrate how to employ PAD with a model-driven app and a canvas app that we'll develop

By applying these concepts and following the examples provided, you will be able to construct PAD flows. This hands-on experience will enable you to gauge the potential of PAD for automating Power Apps testing.

Technical requirements

Access to a functioning Power Platform environment is necessary to follow the example that we will describe in this chapter. The Power Platform developer environment that we explained how to create in *Chapter 5* appears to be the most suitable solution for this purpose.

Additionally, ensure that your system meets the minimum hardware requirements for PAD (`https://learn.microsoft.com/en-us/power-automate/desktop-flows/requirements`).

While a Power Automate trial is not strictly necessary for the features that we will explain in this chapter, if you're interested in exploring more advanced features or capabilities, you may consider signing up for a trial.

We will utilize both a canvas app with a calculator provided in the Test Engine GitHub repository (`https://github.com/microsoft/PowerApps-TestEngine/tree/main/samples/calculator`) and a simple model-driven app created from the `Contacts` table we used in *Chapter 5*, where demo data was imported. For convenience, we have included both solutions in this chapter's GitHub repository (`https://github.com/PacktPublishing/Automate-Testing-for-Power-Apps/tree/main/chapter-08`).

Since both applications are accessible as web applications, we can leverage PAD to perform testing on them.

PAD and business processes testing

PAD allows us to create automated workflows that interact with diverse applications, including web applications. These workflows are designed to execute tasks and validate the user interface, making them particularly valuable for testing Power Apps. Since Power Apps can consist of both model-driven and canvas apps, PAD serves as an additional tool for covering certain parts of end-to-end business processes, similar to a model-driven application.

UI automation with PAD

PAD provides a comprehensive set of UI automation actions specifically designed for interacting with web applications. While we won't extensively cover all the functionalities of PAD, we will outline some of the available actions that are relevant to our purposes, particularly those related to browser interaction.

> **Note**
>
> For more detailed coverage, you can refer to the book *Democratizing RPA with Power Automate Desktop*, by Packt Publishing: `https://www.packtpub.com/product/democratizing-rpa-with-power-automate-desktop/9781803245942`.

The available actions in PAD can be utilized to simulate user interactions within web applications. These actions encompass a wide range of tasks, including clicking buttons, entering text, and navigating through menus. By automating these interactions, you can thoroughly test the functionality and responsiveness of your application's user interface. Additionally, you can assess the integration between various components and systems, ensuring seamless communication and proper synchronization.

To perform UI testing with PAD for web applications, you can create a flow that incorporates the following UI automation actions:

- **Launching a web application**: Utilize the **Launch new Chrome** (or any other browser available) action to initiate the web application you wish to test. In *Figure 8.1*, you can see the actions available, filtered by the `browser` keyword, along with the options available within each action to perform:

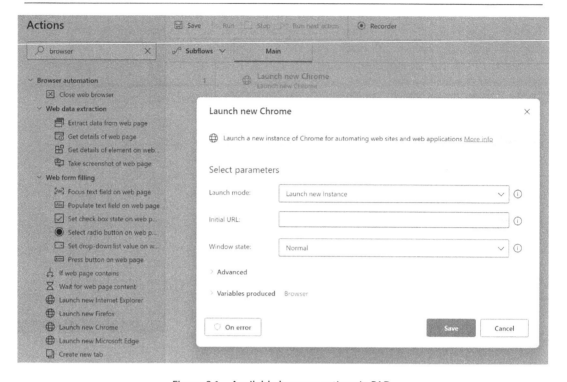

Figure 8.1 – Available browser actions in PAD

- **Interacting with UI elements**: Employ various UI automation actions, such as **Click button**, **Set text**, and **Select item from list** to interact with the web application's user interface. These actions simulate user interactions and allow you to test the functionality and responsiveness of the application.

- **Capturing and validating data**: Implement actions such as **Get text** and **Find element** to capture data from the web application and validate it against expected values. These actions enable you to verify the correctness of data displayed on the web page.

- **Error handling and reporting**: Incorporate error handling mechanisms within your flow to gracefully handle any encountered issues during the testing process. Additionally, implement reporting mechanisms to log and track any errors or exceptions that occur, providing valuable feedback for debugging and analysis.

By structuring your flow with these UI automation actions, you can effectively perform UI testing on web applications using PAD.

For the examples covered in this chapter, we will focus on specific actions that are relevant to testing automation using PAD. While we won't cover all available actions in PAD, the actions discussed will provide you with a solid understanding of the possibilities and capabilities related to testing automation with PAD.

Key considerations while using PAD as a testing tool

UI testing with PAD can be a component of a comprehensive end-to-end testing strategy. This strategy encompasses the entire application flow, from the initial stages to the final user interactions, and covers various user scenarios, interactions, and interfaces. This is the juncture where PAD comes into play, offering coverage for specific areas that may otherwise remain inaccessible with the use of other tools.

For example, end-to-end testing with PAD plays a significant role in validating the functionality and performance of web applications, including model-driven apps. It allows you to replicate real-life settings by leveraging realistic data and user scenarios. By simulating these conditions, you can ensure that your application meets the requirements outlined during its design and development, ultimately delivering a reliable and user-friendly experience.

Best practices

When utilizing PAD for UI testing, it is essential to follow these best practices, which parallel the ones recommended for using other tools.:

- **Identify potential test scenarios**: Identify the critical functionality and user interactions that need to be tested and where PAD can serve as the appropriate tool.

- **Modularize your flows**: Break down your test flows into smaller, reusable components, ensuring not to fragment them excessively. By modularizing your flows, you make them more manageable, maintainable, and easier to update.

- **Implement error handling**: Include error handling mechanisms within your flows to capture and report any issues encountered during testing.

- **Validate test results**: Compare the results of your automated tests against the expected outcomes. This validation step is crucial in ensuring the accuracy and reliability of your test results.

- **Monitor and maintain your flows**: Regularly review and update your test flows to keep them aligned with any changes or updates in your application. As your application evolves, it is essential to ensure that your test flows continue to provide accurate and reliable results.

By following these best practices, you can maximize the effectiveness and efficiency of your UI testing with PAD, leading to robust and reliable application testing.

Benefits of UI testing with PAD

Utilizing PAD for UI testing offers several advantages, many of which are common across other UI testing tools. These include the following:

- **Reduced manual effort**: UI testing automation reduces the time and effort required for manual testing, allowing your team to focus on other important tasks.

- **Increased test coverage**: By leveraging PAD, you can test end-to-end processes that involve interactions with multiple programs or systems. For example, you can verify that the results obtained in a Power Apps app match those in an external tool such as SAP. This allows you to cover more test scenarios in a single automated test.

- **Improved accuracy**: Automated tests are less prone to human error, resulting in more accurate and reliable test results. Since PAD interacts with the actual user interface, it replicates the real-life processes and increases the accuracy of the testing.

- **Faster feedback**: PAD enables you to automate the process of sending test results. Similar to what we did in *Chapter 5* with Test Studio, where we saved results to OneDrive or sent them via email, PAD can automate the treatment of test outputs, sending them by email or saving them in repositories. This can help provide faster feedback on the test outcomes in specific scenarios.

In conclusion, PAD is a valuable tool for UI testing in business processes, particularly for web applications such as Power Apps. By automating user interactions and validating the functionality of your application's user interface and its integration with other components and systems, you can improve software quality and reliability while reducing manual testing efforts.

Now, let's delve into a couple of examples – one with a model-driven app and another with a simple canvas app. These examples will showcase the possibilities of automating tests in Power Apps using RPA, specifically with PAD.

A step-by-step guide to creating a desktop flow for testing a model-driven app and a canvas app

As previously discussed, one of the significant advantages of using PAD for automated testing is its adaptability to nearly any user interface, given that we're simulating user interactions. For this reason, in this section, it will be worth examining how we can tailor the browser action within PAD to test both model-driven apps and canvas apps.

Let's begin by creating a sample model-driven app that utilizes the `Contact` table, where we have already imported some records as per *Chapter 5*. We'll then proceed with the necessary steps to test this app and an example canvas app.

Creating a simple model-driven app

To test the model-driven app aspect, let's begin by creating a basic model-driven app.

To proceed, we will navigate to our environment at `make.powerapps.com` and select the **Tables** tab. Here, we will locate the `Contact` table, which we imported some demonstration data into during *Chapter 5*, as shown in the following screenshot:

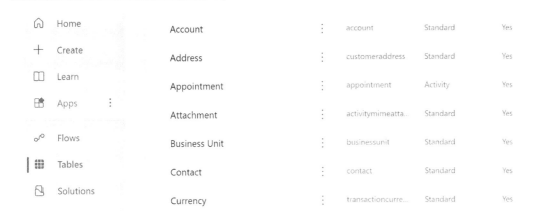

Figure 8.2 – The Contact table with demo data

To access the table details, simply click on **Contact** in the **Table** column. This will take us to the detailed view of the Contact table, as shown here:

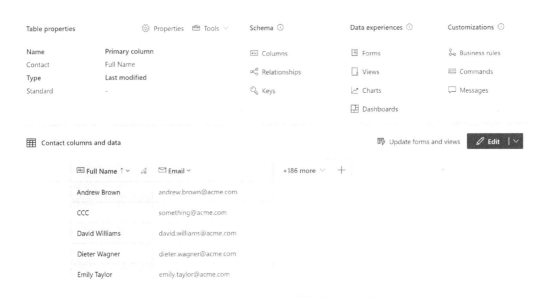

Figure 8.3 – Detailed view of the Contact table

Within the Contact table window, located in the top bar, you will find the **Create an app** button. Click on this button to initiate the app creation process. Let's propose a name for the model-driven app, such as ContactsMDA. The following screenshot shows the initial screen of the app creation process:

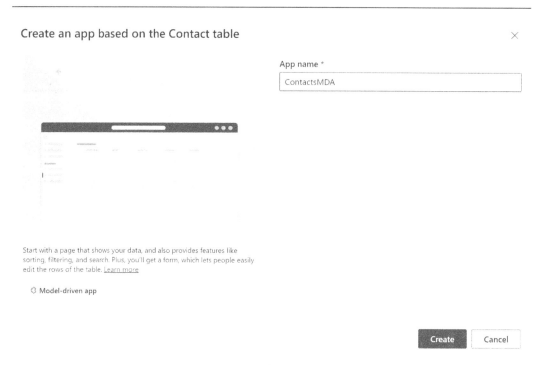

Figure 8.4 – Create an app based on the Contact table

Upon clicking **Create**, you will be able to monitor the progress of the app creation process. This will provide you with a progress message of the creation process, as shown in the following screenshot:

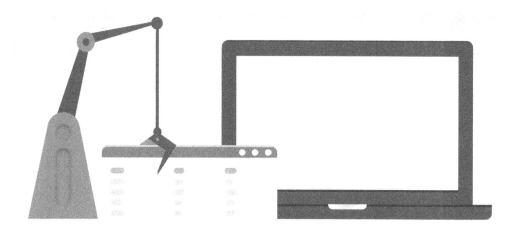

We're using the table to create an app. Next, we'll see it in app designer and take a quick tour. Learn more

Figure 8.5 – App from table creation

As the final step, you will be directed to **App Designer**, where you can view and modify various elements such as forms, views, areas, and more before publishing the app. However, for the test we are conducting, it will be sufficient to publish the application as-is, without making any further modifications, as shown here:

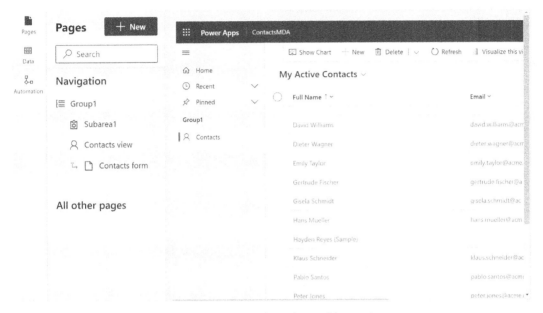

Figure 8.6 – Result of app from table creation

In **App Designer**, locate the **Publish** button and click on it. This action will initiate the publishing process for the app.

Once the publishing process has been completed, you can navigate to the **Apps** tab. There, you will find the application listed. You can either click on the name of the application to open it or click on the three dots located to the right of the name. By selecting the three dots, you will have the option not only to execute the application but also to edit it, share it, and access other options related to app creation, as shown here:

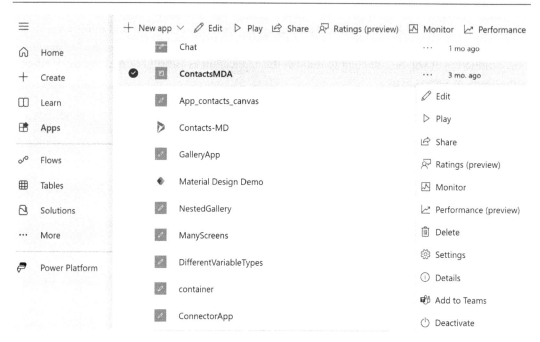

Figure 8.7 – Opening our application

When executing the application, a URL will open in your browser. This URL can be leveraged within our PAD flow for testing purposes. It serves as the link that allows us to interact with and test the application. You can see the initial aspect in the following screenshot:

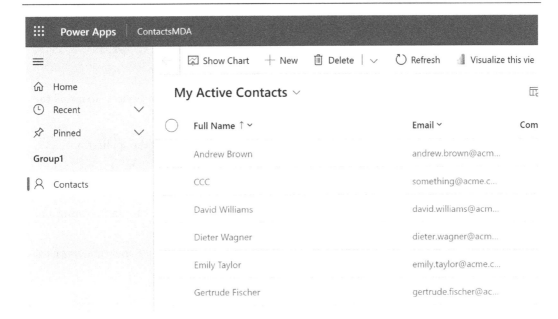

Figure 8.8 – Contacts model-driven app

Now that our model-driven application has been prepared, let's advance to the next section, where we'll create a flow to test it.

Creating a PAD flow to test a model-driven app and a canvas app

Before we proceed with the installation and setup process for PAD on a Windows operating system, it is important to verify that your system meets the requirements outlined in the *Technical requirements* section. Ensure that you have reviewed and met all the specified prerequisites.

Now, let's move on to the steps involved in installing and getting started with PAD.

Downloading PAD from the Microsoft Store

To download PAD from the Microsoft Store, follow these steps:

1. Visit `https://apps.microsoft.com/store/detail/power-automate /9NFTCH6J7FHV`.
2. Click on the **Get** button.
3. The Microsoft Store app should open automatically. If it doesn't, open the Microsoft Store app manually and search for `Power Automate Desktop`.
4. Click on the **Install** button in the Microsoft Store app to start the download and installation process.

Next, we will learn how to install browser extensions.

Installing browser extensions

PAD may require browser extensions to work with specific web applications or services. To install the necessary browser extensions, follow these steps:

1. Launch PAD and click on the **Sign in** button in the top-right corner of the application.

2. Enter your Microsoft account credentials (email and password) and click **Next**.

3. If prompted, enter your Power Platform tenant URL and click **Next**.

4. You should now be connected to your Power Platform tenant and ready to start creating and managing your automation workflows.

5. Select the appropriate developer environment from the top bar if multiple environments exist within the tenant to which you're connecting.

6. In the **Tools** section of the toolbar menu, click **Browser extensions**.

7. Follow the instructions provided to install the browser extensions for your preferred web browsers (such as Google Chrome or Microsoft Edge).

8. Restart PAD for the changes to take effect.

You will be presented with the initial screen in the PAD client, as shown in the following figure:

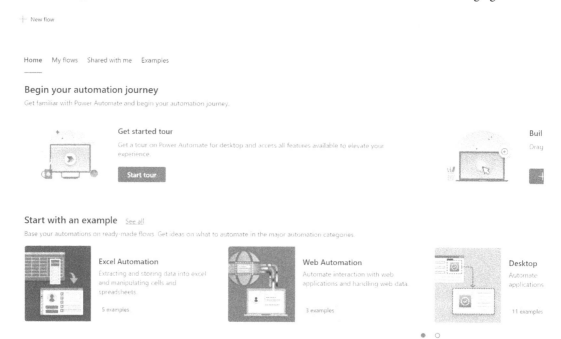

Figure 8.9 – Initial screen in PAD

Next, let's proceed to creating our first PAD flow.

Creating our first PAD flow

Now that your PAD client is ready, let's initiate the flow creation process by clicking on the **New flow** button. An additional window will appear, prompting you to name your PAD flow. You can provide a descriptive name such as `PowerApps Testing` to identify the purpose of the flow:

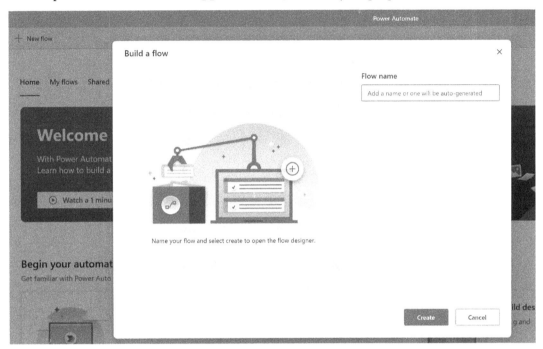

Figure 8.10 – Our first PAD flow

After naming your PAD flow, you will notice a left section displaying all the available components or actions that can be used to create the automation. The central section will contain your main flow, where you can drag and drop actions to build the sequence of steps. On the right-hand side, you will find the variables used in your automation, allowing you to define and manage them and the possibility of moving through UI elements and images to be used in the flow.

While we won't delve into the detailed functionality of these sections in this book, I recommend referring to the comprehensive resource *Democratizing RPA with Power Automate Desktop*, by *Packt Publishing*, which covers these details extensively. It provides a thorough understanding of how to leverage the components and features of PAD to build powerful automations:

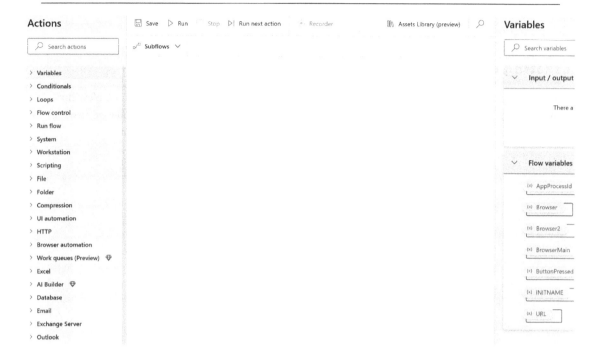

Figure 8.11 – The Actions and Variables sections in a PAD flow

To enable reuse and combination in testing, *we will create separate subflows* for testing the model-driven app and the canvas app. This approach will allow us to modularize our automation and easily reuse and combine the subflows based on the specific case we are testing.

Creating subflows in Power Automate is of significant importance and provides immense convenience. Subflows offer a modular approach to designing automation workflows, allowing users to break down complex processes into smaller, reusable components. This modular design promotes better organization and maintainability, making it easier to manage and update specific parts of the automation.

In our specific case, utilizing subflows would enable us to reuse the login components and easily enable or disable them based on whether we are testing the canvas app or the model-driven app. By separating the common login functionality into a subflow, you can avoid duplication of efforts and ensure consistency in the authentication process.

> **Note**
>
> This approach allows you to create a single main flow that can be shared between the canvas app and the model-driven app. Depending on the scenario, you can activate or deactivate the login subflow, thereby adapting the flow to the specific application being tested.

Let´s start creating a subflow:

1. Navigate to the **Subflows** dialog section and select the option for creating a new subflow. Choose an appropriate name for the subflow, such as `login`.

2. In our new subflow, we will include the **Launch new Chrome** action (or Firefox or Edge, depending on the intended browser). Drag and drop – or double-click – this action into the subflow.

3. In the details of this action, select **Launch new Instance** under **Launch mode**. Set the initial URL to the URL of our model-driven app. Additionally, mark the window state as maximized – the default state is normal. Retain the **Browser** variable that was produced in this step without making any changes:

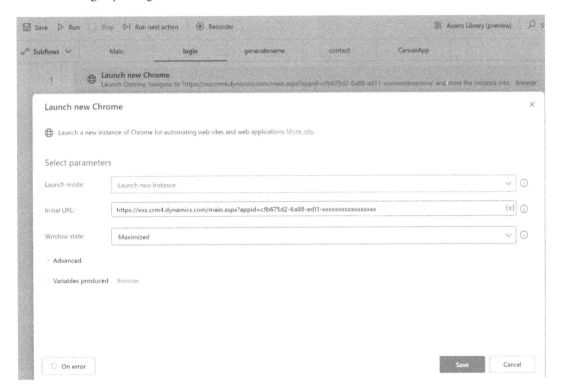

Figure 8.12 – Launch browser action

4. In our main flow, drag and drop the **Run Subflow** action. Select the previously created **login** subflow as the subflow to be executed:

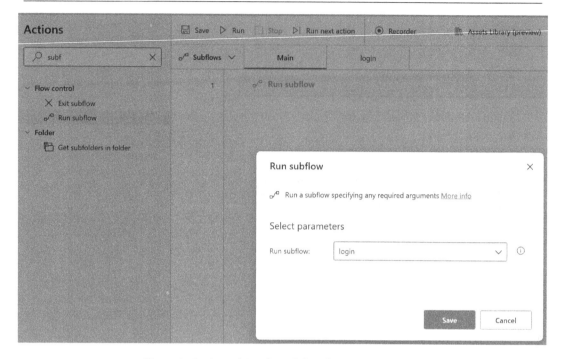

Figure 8.13 – Launching the subflow from the main flow

After saving our progress, we can execute the main flow and observe that it will open the chosen browser. As authentication is required, it will display the login page of our tenant. To complete the login functionality, we will utilize another component called **Recorder**.

Using the Recorder component

To begin with, in PAD, and with our **login** subflow open, click on the **Recorder** button; this will open the recorder on top of our browser. Once opened, follow these steps:

1.   Click **Record** and start recording the actions related to the login process:

   i.   Fill in the account login details.

   ii.   Click on **Next** and enter the password:

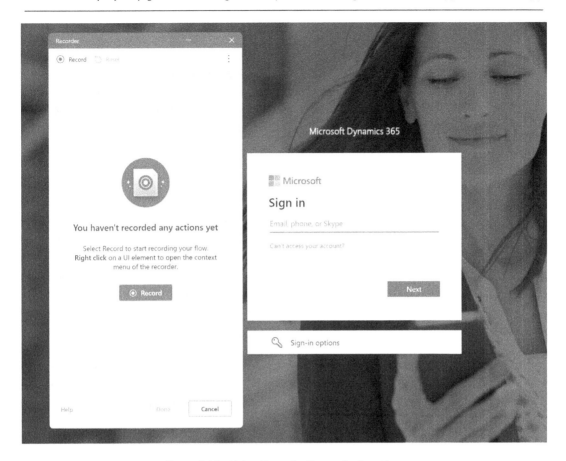

Figure 8.14 – Using Recorder for our login actions

2. Next, we will need to confirm that we wish to stay signed in, after which the screen showing the **Contacts** model-driven app will be presented.

    We can pause the **Recorder** component, check the actions that we have created, and remove any actions that were recorded by mistake.

3. Once we are satisfied with the final result, click on **Done** in the **Recorder** component and add the actions to our **login** subflow:

Figure 8.15 – Actions created by the Recorder component

> **Note**
>
> The necessity and complexity of a login subflow can vary significantly depending on the authentication policies of your environment. In some cases, it may not be required if your organization has **single sign-on** (**SSO**) implemented. On the other hand, if **multi-factor authentication** (**MFA**) is in use, you might need to construct a workaround. This can be achieved using conditional access policies for the testing use (`https://learn.microsoft.com/en-us/troubleshoot/power-platform/power-automate/conditional-access-and-multi-factor-authentication-in-flow`).

We're currently logged in to our flow, but the next step is to automate the desired actions. Specifically, we will create a new contact within the application.

Testing the contact creation process

Now that we are on the initial page of our application, let's bring the **Recorder** component to the forefront again:

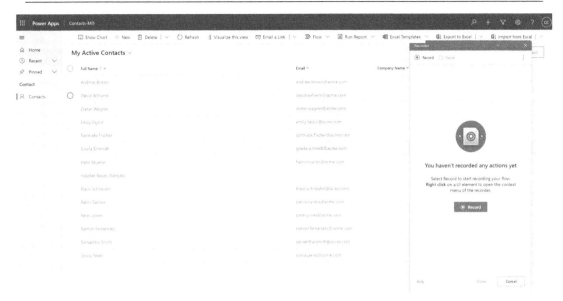

Figure 8.16 – Using the Recorder component for contact creation actions

Follow these steps:

1. To better organize our flow, we will create a new subflow called `contact`. This will allow us to activate or deactivate this subflow and test different possibilities, such as testing the canvas app or the model-driven app.

2. Click on **Record** to begin recording and perform the actions that we want to include in the test.

   These actions include the following:

   i.   Click on the **New** button in the contacts screen.

   ii. Fill in the mandatory fields, which in this case are the **Last Name** field and any optional field that we want to test (such as the **Email** field).

   iii. Click **Save** in the model-driven app to save the contact.

We will observe that the recorder has captured these actions:

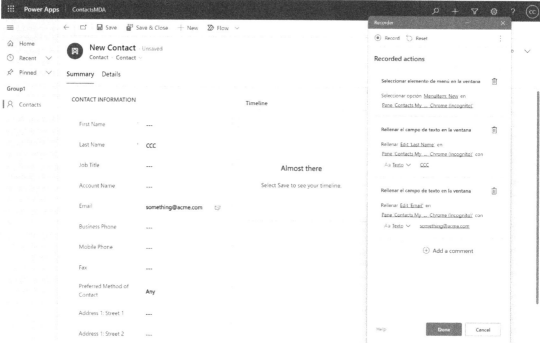

Figure 8.17 – Contact creation recorded

3.   After clicking **Done**, we will see that these actions have been added to our PAD subflow called **contact**:

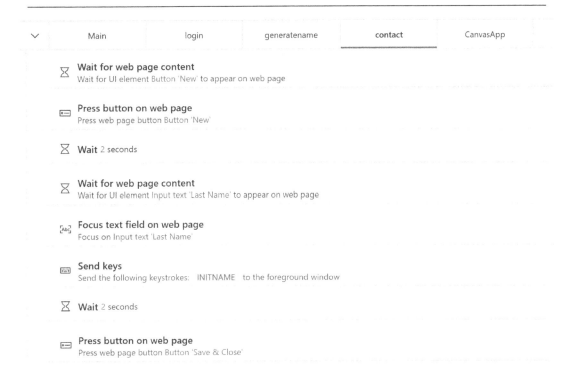

Figure 8.18 – Contact creation subflow

Now that we have created the **contact** subflow, we can add it to our main flow and proceed to test the application.

However, *we may encounter a message indicating that a duplicate record* was found due to the duplicate record detection rules since we are using the same last name that was used in our recording. In the next section, we will learn how to avoid this message.

Generating a random name for the contact name

To avoid the duplicate record message in our tests, we are going to create a random name using the **Create Random Text** action available in PAD.

In the **Actions** section on the left-hand side of PAD, search for random to find this action. We have the option to create a new subflow for producing this random text, which we can then use to test our canvas app as well – alternatively, you can choose to directly incorporate the action into the main flow. We have called this subflow generatename.

We will drag the **Create Random Text** action to our new subflow and specify what we want to use for the random text creation, such as letters, digits, symbols, and the length of the text. We can store the result of this action in a variable that I have named initname:

> **Note**
>
> Remember to edit the name by only editing the text between the percentage symbols (%) while keeping them in place.

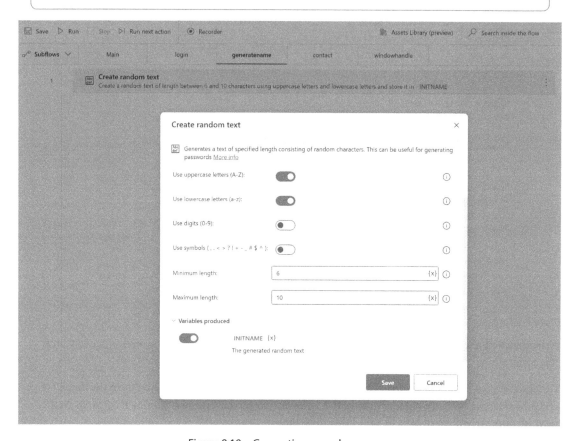

Figure 8.19 – Generating a random name

Now, we can use this subflow in the main flow before the execution of the **Contact Creation** flow. This way, we can use the outcome of this action as the input for the **Last Name** field in the **Contact Creation** subflow:

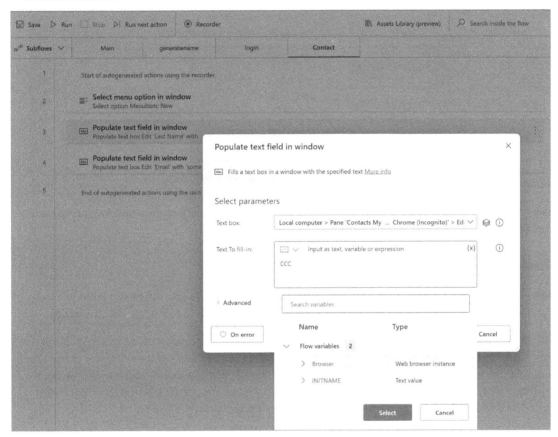

Figure 8.20 – Using the variable generated as a random name

The final order of the subflows in the main flow will be as follows:

⊞ Save	▷ Run	⎵ Stop	▷| Run next action	◉ Recorder

⊡ Subflows ⌄	**Main**	login	generatename	contact

1   ⊡ **Run subflow** generatename

2   ⊡ **Run subflow** login

3   ⊡ **Run subflow** contact

Figure 8.21 – Order of the subflows in the main flow

Now that we have followed the steps to test the model-driven app, let's explore how we can use the same pieces to test a canvas app. The process will be very similar.

Testing a canvas app

Now, let's take a similar approach to test a canvas app. Since the canvas app will be displayed in a web interface for testing, the process will be very similar to the one we used for the model-driven app.

To try it out, let's use one of the canvas apps available in the Test Engine repository (`https://github.com/microsoft/PowerApps-TestEngine`). In this case, we can use the calculator app as an example. As noted in the *Technical requirements* section, the solution that contains the application is also available in this chapter's repository (`https://github.com/PacktPublishing/Automate-Testing-for-Power-Apps/tree/main/chapter-08`):

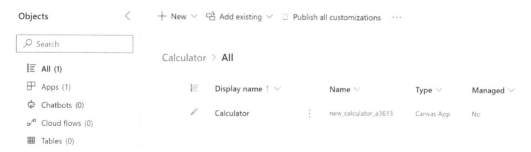

Figure 8.22 – The calculator app available in the Test Engine repository

The calculator app will present us with a simple interface featuring two number fields (both with a value of 100) and the available operations, such as addition, multiplication, division, and subtraction, between them:

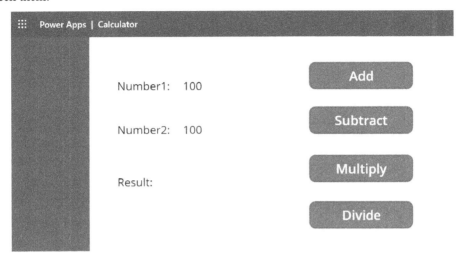

Figure 8.23 – Calculator app interface

For our test, we will automate one of these operations (*addition*) and check that the presented result is correct.

Going back to PAD, we can create a new subflow called CanvasApp and from here launch a new browser instance to navigate to the URL of the app. In this example, I am using Firefox, but you can choose Edge or Chrome.

You can use the **Recorder** component from here and click on the **Add** button of the application:

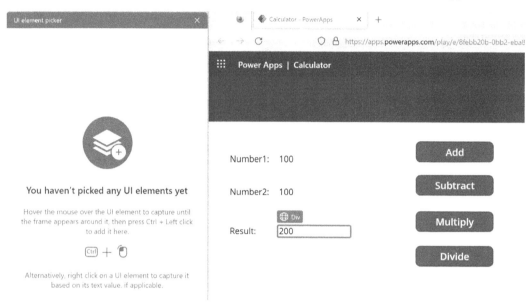

Figure 8.24 – Using the Recorder component to test the canvas

Next, we will need to manually create an action to wait for a UI element to appear (wait for web page content). In the UI element dialog, we can create a new UI element by using the **Add UI element** option and then leveraging the UI element picker by selecting the element we are looking for by using *Ctrl* and left-clicking.

In this case, when we select the result label, we will be selecting a DIV layer called 200 (the name comes from the expected result of the operation), and this UI element will contain the result that we are expecting. Therefore, if we can find this element, the test will have passed correctly:

Subflows ∨	Main	login	generatename	contact	CanvasApp

1   **Launch new Firefox**
Launch Firefox, navigate to 'https://apps.powerapps.com/play/e/
                         _source=portal' and store the instance into Browser2   .?tenantId=

2   Start of autogenerated actions using the recorder

3   **Launch new Firefox**
Attach to Firefox tab with URL 'https://apps.powerapps.com/play/e,'
                    '&source=portal' and store the instance into Browser   u/?tenantId

4   **Click link on web page**
Click on Div 'Add' of web page

5   End of autogenerated actions using the recorder

6   **Wait for web page content**
Wait for UI element Div '200' to appear on web page

7   **Display message**
Display message 'Result is 200 as expected' in the notification popup window with title 'Result OK' and store the button pressed into ButtonPressed

Figure 8.25 – Adding actions to the subflow to check the correct execution

Once we have checked that the expected result is there, we will have the option to display a pop-up message stating that the test was executed correctly, as shown here:

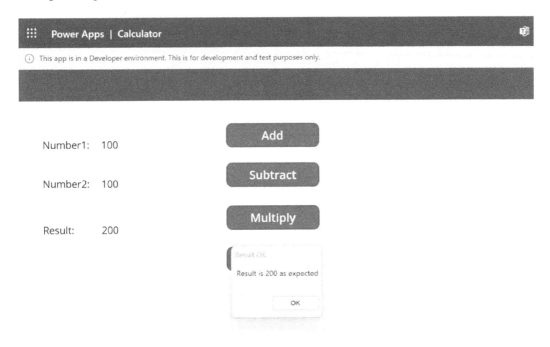

Figure 8.26 – Confirming the correct execution of the operation with a message

We've simplified the execution check using a message box. However, in real-world scenarios, this could be substituted with more complex actions such as sending an email or generating a log in a text file, among other options available in PAD.

Summary

PAD can be a valuable tool for the automated testing of Power Platform in specific scenarios, thereby enhancing test coverage, for example facilitating end-to-end business process testing that involves interaction with model-driven apps and other elements of Power Platform.

While PAD brings benefits such as reduced manual effort and increased accuracy, it may not fit every testing scenario. The effectiveness of PAD depends on factors such as the application type, integration needs, and overall testing strategy. Evaluating these considerations is key before adopting PAD. In appropriate situations, PAD can augment manual testing, providing another option for automated UI testing.

Overall, this chapter aimed to introduce the capabilities of PAD for automated testing, providing a foundation for you to assess its potential role based on your specific testing needs. With the ability to reduce repetitive tasks and increase test coverage, PAD can serve as a useful addition to any Power Apps tester's toolkit in applicable contexts.

Part 3:
Extending Power Apps
Automated Testing

In the last part of the book, we will broaden our exploration to more advanced areas of Power Apps testing. First, we will dive into testing PCF and canvas components with Test Engine, providing practical examples for each. Then, we will examine **Application Lifecycle Management (ALM)** using Test Studio within Azure DevOps pipelines.

We will then introduce the concept of mocks with Test Engine, a powerful testing strategy that imitates app dependencies for more robust testing. Finally, we will delve into the crucial field of telemetry within Power Apps, exploring how analytics can help improve testing and debugging processes.

This part has the following chapters:

- *Chapter 9, PCF and Canvas Components Testing*
- *Chapter 10, ALM with Test Studio and Test Engine*
- *Chapter 11, Mocks with Test Engine*
- *Chapter 12, Telemetry and Power Apps*

9

PCF and Canvas Component Testing

In the previous chapters, we embarked on a journey through the fundamentals of **Test Engine**, exploring its capabilities, benefits, and how to effectively use it to test Power Apps. We covered everything from importing sample solutions and executing test plans to authoring new test plans in YAML and reusing existing **Test Studio** assets.

In this chapter, we'll build on that foundation and delve into a significant advantage of Test Engine over Test Studio: the ability to automate the testing of Canvas components and **Power Apps Component Framework** (PCF) components. These components are powerful tools for customizing and enhancing your Power Apps; being able to automate their testing is a significant benefit for ensuring the reliability and functionality of your applications.

First, we'll provide an overview of Canvas components and walk you through a practical example of their use. Next, we'll do the same for PCF components, giving you a solid understanding of these essential building blocks in Power Apps development.

Once you've grasped the basics of these components, we'll guide you through the process of authoring tests for Canvas components and PCF components in Test Engine. You'll learn how to create these tests, run them using Test Engine, and inspect the results in the output folder.

In this chapter, we will cover the following topics:

- **Exploring Canvas components**: This section will provide a concise explanation of Canvas components, along with a practical application example for better comprehension
- **Exploring PCF components**: In this section, you will gain insights into PCF components and learn their effective usage

- **Testing Canvas components**: This section will equip you with the skills necessary to create tests for Canvas components and execute them using Test Engine

- **Testing PCF components**: In this section, you will learn how to create tests for PCF components, run these tests via Test Engine, and subsequently review the outcomes

Armed with these skills, you'll be ready to take full advantage of Test Engine's capabilities for testing Canvas and PCF components in your Power Apps. As we move forward, we'll continue to provide you with tips, tricks, and best practices to make your Power Apps testing as effective and efficient as possible. Let's get started!

Technical requirements

Before we delve into the practical examples of this chapter, it's crucial to ensure that certain elements, as established in previous chapters, are in place:

- **Test Engine executable**: This should be already constructed on your local machine. The process of installing prerequisites, including the .NET Core SDK, and then compiling the Test Engine source code from GitHub, was covered in *Chapter 6*. Alternatively, you can use the new Power Platform CLI (**pac cli**), which we described in *Chapter 7*.

- **Sample applications**: The sample application from the Test Engine repository (PCFcomponent_1_0_0_3.zip) should be integrated into your Power Apps environment for PCF testing. This chapter also includes the procedure for creating a Canvas component before testing it. However, as an alternative, you can choose to download the sample solution from the corresponding chapter in this book's GitHub repository.

> **Note**
>
> This chapter's GitHub repository includes both the Canvas component solution that we are creating and the PCF component solution from the Test Engine repository (https://github.com/PacktPublishing/Automate-Testing-for-Power-Apps/tree/main/chapter-09).

Please verify that Test Engine has been compiled and set up correctly before advancing with the examples in this chapter. If it's not, please revisit *Chapter 6* to establish Test Engine on your system and import the samples.

With the samples prepared in your environment and Test Engine configured, you're equipped to start authoring and executing test plans. As we navigate through the examples in this chapter, you'll get to witness Test Engine in action with Canvas and PCF components. Now, let's dive in!

Exploring Canvas components and their application in a practical example

Canvas components are reusable parts or building blocks in Power Apps that can be used across different screens within an app. They promote *code reusability and help maintain consistency* across your application. A Canvas component could be something as simple as a button that's used in multiple places, or as complex as a custom calendar control.

Let's understand these components better through a practical example. Imagine you're building an application where you have a specific layout of labels, text inputs, and buttons that need to be repeated on multiple screens. Instead of manually adding and configuring these controls on each screen, you can create a canvas component that defines this layout.

Now, whenever you need this particular layout of controls, you can simply insert the component into your screen rather than adding each control individually. You can do this by going to the **Insert** tab in the left navigation bar, choosing **Get more components**, and then selecting the component you created.

The real power of Canvas components comes from their ability to define *custom properties*. These properties allow you to customize the behavior and appearance of the component for each instance of it in your app. For example, you could define a custom property that controls the text of `Label1`. Then, whenever you insert the component into a screen, you can set this property to whatever text you want `Label1` to display on that screen.

By using Canvas components, you streamline your app development and make your app easier to maintain and update. Before we test a canvas component, let´s create a new one for our purpose.

Creating a UserCard component

This section will walk you through the process of building a `UserCard` component in Power Apps. This multifunctional and reusable component encapsulates key user details and functionalities, including a user's profile image, as well as their name, role, and status, and a button to initiate chat. This component could enhance the efficiency of our app development process by being useful in various contexts within an app, such as in a staff directory, a contact list, or a chat app. For our testing purposes, we will be adding a status label to simulate the status field in a chat application.

Enabling enhanced component properties

Before we can use the **Enhanced Component Properties** setting of our canvas app, we have to enable the new functionality. Follow these steps to do so:

1. Navigate to your Power Apps settings.
2. Go to **Upcoming features** | **Experimental**.
3. Activate the feature.
4. Save the app and reload the Power Apps designer.

Building the UserCard component

As the first step, we will be creating a new component. Follow these steps to do so:

1. Open Power Apps Studio.

2. In the left-hand navigation pane, click **Tree View** and then choose the **Components** tab to open the **Components** screen.

3. Click on the **+ New component** button at the top to create a new blank component.

4. You will see a new blank screen that represents your component. Rename your component UserCard at the top right-hand side of the page. The following figure shows an example of this:

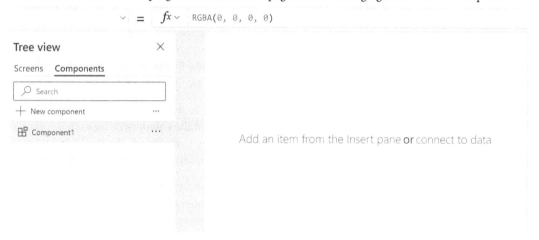

Figure 9.1 – Creating a new Canvas component

Next, we will be adding controls to the component.

We need to add a circle and an image for the profile picture, two labels for the user's name and role, a button for initiating a chat, and a label for the user's status.

Here's how to do it:

1. To add the circle, go to the **Insert** tab, select the **Shapes** dropdown, and choose **Circle**. This will serve as a placeholder for the user's profile picture. Insert an **Image** element here.

2. To add the labels, go to the **Insert** tab and select **Label**. Do this twice to create two labels: one for the user's name (let's call it lblUserName) and another for the user's role (let's call it lblUserRole).

3. To add the chat button, go to the **Insert** tab and select the **btnChat** button. This will be our chat initiation button, as we will describe later.

4. To add the status label, go to the **Insert** tab and select **Label**. This will create a label for the user's status (let's call it `lblUserStatus`).

The following figure depicts a similar implementation:

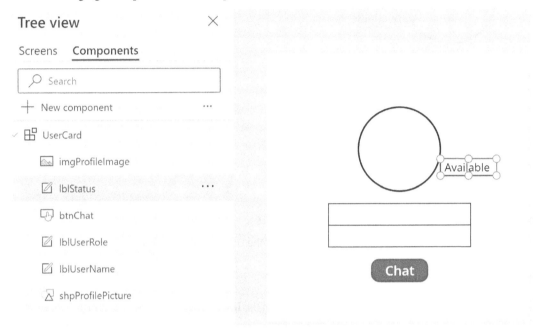

Figure 9.2 – The UserCard component

The next step will be defining **Custom Properties** for the controls:

1. Select the component in the **Components** screen, click on the **Custom Properties** tab in the right-hand pane, and then click + **Add a custom property**. Name the property `ProfilePicture` and set it as an **Image** type.

2. Repeat this process to create `UserName` and `UserRole` properties of the **Text** type.

3. To create a custom event property for the button, select the component in the tree view and click on the + icon to create a new custom property. Choose the **Events** property type, name it `OnSelectChat`, and click **Create**.

4. Create a final custom property named `Status`. Set it as a **Text** type. This property will control the text of **UserStatusLabel**. The following figure showcases an example configuration:

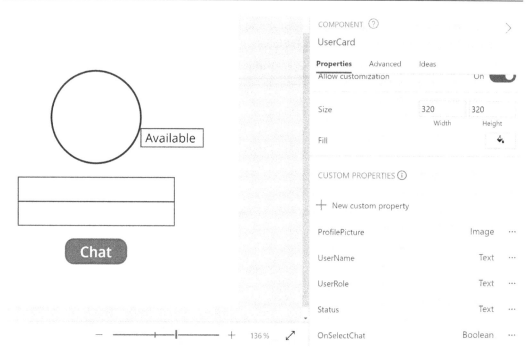

Figure 9.3 – Custom properties of the UserCard component

Now, we will link **Properties** to **Controls**:

1. Select the image, go to the **Properties** pane, and set the **Image** property to **UserCard.ProfilePicture**.
2. Select **UserNameLabel**, go to the **Properties** pane, and set the **Text** property to **UserCard.UserName**.
3. Select **UserRoleLabel**, go to the **Properties** pane, and set the **Text** property to **UserCard.UserRole**.
4. Select **UserStatusLabel**, go to the **Properties** pane, and set the **Text** property to **UserCard.Status**.
5. Select the button, go to the **OnSelect** property, and set it to **UserCard.OnSelectChat()**.

    This will link the **OnSelect** button event with this property, which can be defined in the host application of the component.

> **Note**
>
> This custom property is the most important one to link for our use case as we'll be using it for our test.

The following figure shows this configuration:

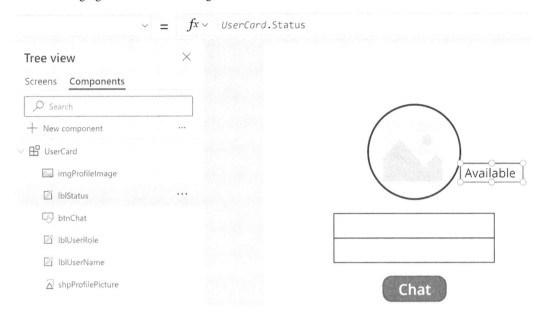

Figure 9.4 – The custom property linked to UserStatusLabel

Now that the custom property has been linked, we will arrange and style the controls:

1. Arrange and style the controls as desired.

    For example, you can place the circle at the top, the labels beneath it, and the button at the bottom.

2. Save your work. Once you're happy with your component, click **Save** to save your work and click **Publish**.

Now, your UserCard component is ready for use. Whenever you add an instance of this component to a screen in your app, you can set the custom properties to control the profile picture, the user's name, the user's role, the action when the button is clicked, and the user's status.

Utilizing the UserCard component

To utilize the UserCard component in your Power Apps canvas app, adhere to the following steps:

1. Navigate to the desired screen. Within Power Apps Studio, navigate to the screen where you wish to integrate the UserCard component. Proceed to the **Insert** tab, select **Library Components**, and then choose the UserCard component from the list (if you are using the component in the same app where you have created it, you can utilize **Insert** and **Custom** from the top bar).

If the component is being used within the same application where it was initially created, it should be readily available. If not, navigate to **Get more components** and import the component library that you've created. The available options are shown in the following screenshot:

Figure 9.5 – Using a Canvas component in the application

2. Set **Custom Properties**. After the UserCard component has been added to the screen, its custom properties can be set in the **Properties** panel. For instance, you can set the ProfilePicture, UserName, UserRole, and UserStatus properties to exhibit the desired user information. These properties can be defined with a variable, the value of which can be modified at runtime during the application's execution. Check the initial values that have been applied for these properties:

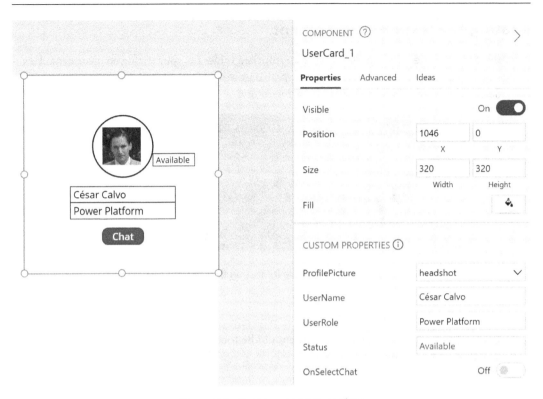

Figure 9.6 – Custom properties values

3. Define the OnSelectChat event. Within the **Properties** panel, locate the custom **OnSelectChat** event property you created previously. Here, you can define the logic to be executed when the chat button is clicked. This could include navigating to a different screen or updating a variable.

4. Repeat this process for multiple instances. Should you need to present multiple **UserCard** components on the same screen, additional instances of the component can be added with their custom properties set accordingly.

Once you've configured the UserCard components, save your app and publish it. In the next section, we'll test it by altering the component properties using Test Engine.

Testing canvas components with Test Engine

To conduct testing on the UserCard component using Test Engine, we must first configure the testing environment.

Updating the testing environment

Modify the `config.dev.json` file with the appropriate values for your testing environment. This includes `environmentId`, `tenantId`, `outputDirectory`, and, of course, the path to the YAML file containing the test suite, as shown in *Figure 9.7*:

```
te3 > PowerApps-TestEngine > src > PowerAppsTestEngine > {} config.dev.json > ...
    {
        "environmentId": "8febb20b-0bb2-eba8-ae06-c5790023ca49",
        "tenantId": "be66fb28-a88d-47ff-8838-1f33b8d0bbbf",
        "testPlanFile": "c:/te3/PowerApps-TestEngine/samples/basicgallery/canvascomponent.fx.yaml",
        "outputDirectory": "D:/results/canvascomponent",
        "logLevel": "",
        "domain":  "",
        "queryParams": ""
    }
```

Figure 9.7 – The config.dev.json file for testing the Canvas component

> **Note**
>
> Remember, the `config.dev.json` file should be located in the same folder where the `TestEngine` command is being executed.

Creating a test suite

Construct a test suite for the `UserCard` component. This suite will include a test case for altering the UserCard's status and `userrole` properties.

The goal is to demonstrate that, with `TestEngine`, we can test canvas components. We can apply this pattern in a canvas app that includes a canvas component, thereby combining the testing of controls with the testing of canvas components. The following figure suggests how to implement this test case:

```
1   testSuite:
2     testSuiteName: CanvasComponent
3     testSuiteDescription: ''
4     persona: User1
5     appLogicalName: cccatp_chat_06d07
6     appId: ''
7     onTestCaseStart: ""
8     onTestCaseComplete: ""
9     onTestSuiteComplete: ""
10    networkRequestMocks:
11    testCases:
12    - testCaseName: Check Chat button
13      testCaseDescription: ''
14      testSteps: |
15        =
16        Screenshot("UserCard_loaded.png");
17        SetProperty(UserCard_1.Status, "Test");
18        Assert(UserCard_1.Status = "Test", "Make sure status is set to Test");
19        Screenshot("Usercard_statuschanged.png");
20        SetProperty(UserCard_1.UserRole, "Developer");
21        Assert(UserCard_1.UserRole = "Developer", "Make sure role is set to Developer");
22        Screenshot("Usercard_end.png");
23
24    testSettings:
25      locale: "en-US"
26      recordVideo: true
27      headless: false
28      browserConfigurations:
29        - browser: Chromium
30    environmentVariables:
31      users:
32        - personaName: User1
33          emailKey: user1Email
34          passwordKey: user1Password
35
```

Figure 9.8 – The YAML file to test the UserCard Canvas component

Start by specifying the *name of the application to test* – you can use either the logical name or AppId. Here, we're using the logical name as the application is included in a solution that is available in our GitHub repository.

We will continue with the same *authentication configuration* we used in *Chapter 7*, using the User1 persona and the authentication values saved in environment variables.

Set the value of **headless** to **false** to monitor the browser interactions during the test, and **recordVideo** to **true** to review the video later.

Lastly, we must configure the test case:

1. In `testSteps`, create a screenshot at the start of the test.

2. Set a value for the user's status by modifying the `component` property.

3. Assert the value change to validate this part of the test.

4. Take another screenshot for later graphical reference.

5. Set a value for the user role by altering the `component` property.

6. Assert the value change to validate this part of the test.

Finally, take another screenshot of the final screen to verify that both values have been modified.

Running the test suite

Execute the test suite using the Test Engine CLI in the folder where you downloaded the TestEngine GitHub project, which contains the `src` and `PowerAppsTestEngine` folders – for example, you can run `..\PowerApps-TestEngine\src\PowerAppsTestEngine`. This is where we should place our `config.dev.json` file.

Execute the `dotnet run` command, as shown in the following screenshot:

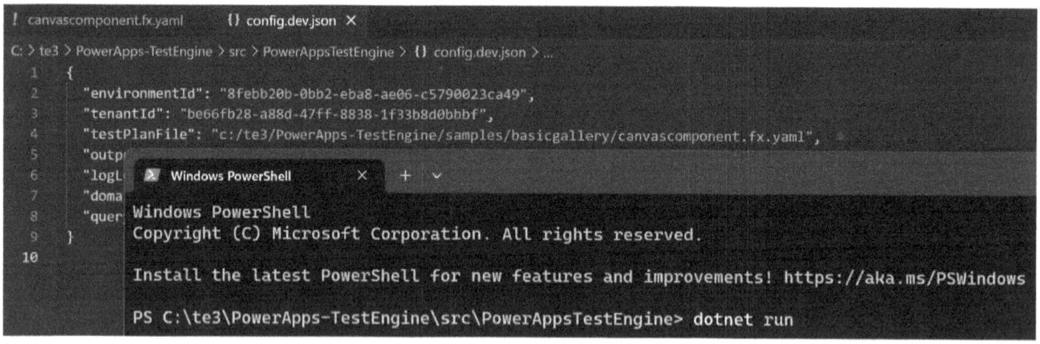

Figure 9.9 – Executing dotnet run to test the canvas component

Test Engine will execute the test steps defined in the YAML file, such as taking screenshots, setting the `UserCard` status property, asserting that the status is set to **Test**, as well as setting the `UserRole` property and asserting its value.

If everything proceeds as expected, you will see a summary similar to the following:

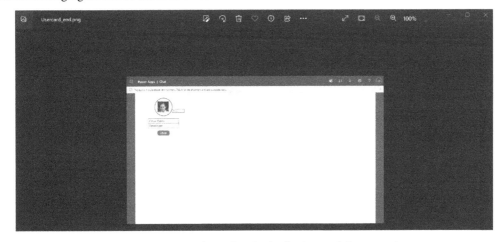

Figure 9.10 – Summary of the canvas component test

This summary provides an overview of the count of test cases executed, the number of tests that passed, and those that failed. Additionally, the summary indicates the destination folder where the test results and `.trx` files are stored. This allows you to easily review the test suite's performance and locate the detailed results for further analysis.

Reviewing the test results

Navigate to the test results folder. Here, you'll find a structure similar to the one we observed in *Chapter 7*, with a root folder dedicated to the execution and storing the `global.trx` file, a subfolder dedicated to the test suite containing the video recording and the logs for the test suite, and another subfolder within this one dedicated to each test case with logs and screenshots concerning these test cases.

The following figure shows the final screenshot we've taken:

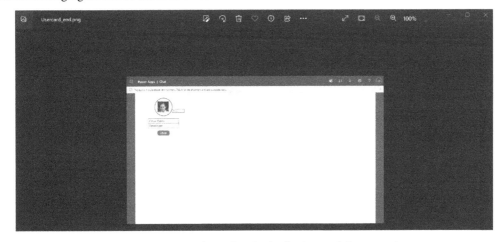

Figure 9.11 – Screenshot taken in the final step of the execution

Now that we've reviewed how to use Test Engine with Canvas components, we can apply a similar approach to test another type of component that's critical in Power Apps development: a PCF component. But before that, let's start by understanding what PCF components are.

Exploring PCF components and how to use them

PCF is a robust framework that empowers professional developers and app makers to craft custom code components for model-driven and canvas apps. These components substantially augment user experiences when interacting with data on forms, views, dashboards, and canvas app screens. The significance of PCF components is grounded in their *ability to extend Power Apps' capabilities* beyond the native components, thereby offering app developers greater flexibility and customization options.

Initially targeted at model-driven apps, PCF components enabled developers to create custom controls for forms, views, and dashboards. However, as time has passed, the integration of PCF components with canvas apps has broadened, *allowing developers to utilize this powerful framework for both Power Apps types and Power Pages*.

The potential for extending Power Apps with PCF components is vast. Developers can generate components using TypeScript, JavaScript, HTML, and CSS, and integrate them with renowned client frameworks such as React and Angular. This opens the door to creating highly interactive and visually engaging components that can be reused across multiple apps, thereby enhancing the user experience and functionality overall.

PCF components can range from custom visualizations and interactive controls to advanced data manipulation tools. These can either replace or enhance existing Power Apps components, leading to more tailored and efficient solutions for specific business requirements.

The integration of PCF components in canvas apps continues to evolve, with consistent improvements and updates to the framework. Consequently, developers can look forward to better support, more features, and increased compatibility with various client frameworks in the future.

Using PCF components in canvas apps

This book does not cover the process of creating a PCF control. For more detailed information on this topic, refer to the book *Extending Microsoft Power Apps with Power Apps Component Framework*, by *Danish Naglekar*, available at *Packt Publishing* (`https://www.packtpub.com/product/extending-microsoft-power-apps-with-power-apps-component-framework/9781800564916`).

In this section, we will delve into *how to use PCF components in a canvas app*, paying specific attention to those provided in the Test Engine GitHub repository for our test sample.

Follow these steps to use PCF components in a canvas app:

1. **Import the component**: Begin by importing the solution with the component you wish to use. We're using an example provided by the Test Engine GitHub repository, specifically the `pcfcomponent` solution, which you can find in the `samples` folder or within this book's repository (named `PCFcomponent_1_0_0_3.zip`). However, feel free to experiment with your own PCF components.

    Figure 9.12 illustrates the PCF component alongside the sample application. Both of these can be located in the sample solution within the Test Engine GitHub repository:

Figure 9.12 – PCF component solution imported

> **Note**
>
> Visit the PCF Gallery, which is developed and maintained by Guido Preite at `https://pcf.gallery/`. Here, you can search for and download PCF components for your Power Apps projects. The website offers a variety of controls created with the Power Apps component framework, which can be filtered by parameters such as canvas components or model-driven apps. You'll find controls for various functionalities, such as file uploaders, iframes, and progress bars. Start experimenting with these PCF components to enhance your Power Apps development experience.

2. **Import a PCF component**: In Power Apps Studio, select **Insert** | **Get More Components**. In the **Import components** dialog, select **Code**. The following screenshot shows an example of a list of PCF components that are readily available for you to import into your application:

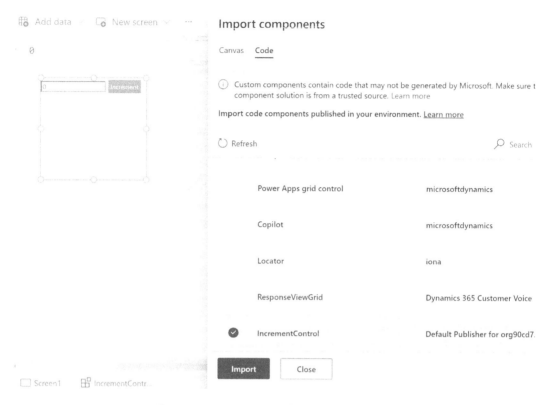

Figure 9.13 – Importing code components (PCF)

> **Note**
>
> You may need to verify your environment settings in the **Admin Center** area. In particular, under the **Features** section, ensure that **Power Apps Component Framework for Canvas Apps** is enabled if it isn't by default.

3. **Insert the PCF component**: After importing the `IncrementControl` component, navigate to the screen where you want to add the component. Go to the **Insert** tab, click on **Code components**, and select the imported PCF component from the list.

4. **Configure the PCF component**: Once the PCF component has been added to the screen, you can configure its properties and behavior using the **Properties** panel. *Figure 9.14* shows the **Properties** panel for our example:

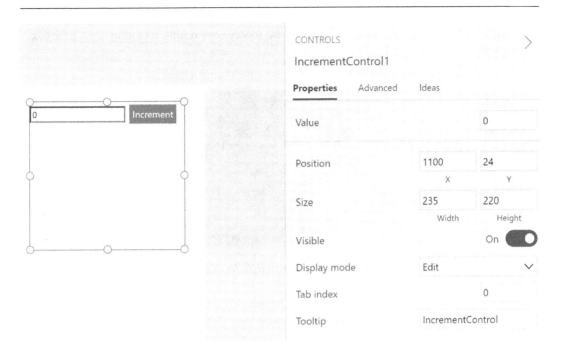

Figure 9.14 – PCF component properties

Now that we have added our PCF component, it´s time to begin the process of testing this PCF component in the new section.

Testing PCF components with Test Engine

Similar to our approach in the Testing canvas components with Test Engine section, we will employ a YAML file capable of modifying the component's properties. This strategy, when coupled with the ability to interact with other controls, allows us to conduct a comprehensive end-to-end test of the application we aim to evaluate.

Our initial step typically involves configuring the `config.dev.json` file. Here's how:

1. Configure the environment ID and tenant ID where the application is situated.

2. Point to the path where our YAML file can be found. In this case, since we're directly using the provided sample file, we must point to the `samples` folder and the accompanying YAML file.

Finally, specify the output folder where we wish to store our results, including the `.trx` file, logs, videos, and screenshots.

Figure 9.15 shows how the `config.dev.json` file should appear:

```
C: > te3 > PowerApps-TestEngine > src > PowerAppsTestEngine > {} config.dev.json > ⋯ outputDirectory
  1    {
  2        "environmentId": "8febb20b-0bb2-eba8-ae06-c5790023ca49",
  3        "tenantId": "be66fb28-a88d-47ff-8838-1f33b8d0bbbf",
  4        "testPlanFile": "C:/te3/PowerApps-TestEngine/samples/pcfcomponent/testPlan.fx.yaml",
  5        "outputDirectory": "D:/results/pcfcomponent",
  6        "logLevel": "",
  7        "domain": "",
  8        "queryParams": ""
  9    }
 10
```

Figure 9.15 – The config.dev.json file for PCF testing

The second step involves the actual YAML file. We're referring directly to the sample file housed in Test Engine's GitHub repository. Here, you can examine the following aspects:

- The test suite's name, description, persona, and `appLogicalName` are already indicated. If we're using the sample solution, there should be no need for alterations here as `appLogicalName` is derived from this solution.

- In the `testCases` section, you'll find a single case (named `Case1`) and the steps that have been incorporated into this test case:

    I.   We start by taking a screenshot of the application at startup.

    II.   The only property available in our control (`incrementControl1`) is set to 10, and we verify that this value is accurately reflected.

    III. We repeat this process with a value of `20`.

    IV.   A final screenshot is taken to capture the results.

Figure 9.16 demonstrates these steps alongside the remaining configuration:

```
C: > te3 > PowerApps-TestEngine > samples > pcfcomponent >  ! testPlan.fx.yaml
 1   testSuite:
 2     testSuiteName: PCF Component
 3     testSuiteDescription: Verifies that you can interact with increment control of the PCF Component
 4     persona: User1
 5     appLogicalName: new_pcfcomponent_bcc03
 6
 7     testCases:
 8       - testCaseName: Case1
 9         testSteps: |
10           = Screenshot("pcfcomponent_loaded.png");
11             SetProperty(IncrementControl1.value, 10);
12             Assert(IncrementControl1.value = 10, "Make sure increment control is set to 10");
13             SetProperty(IncrementControl1.value, 20);
14             Assert(IncrementControl1.value = 20, "Make sure increment control is set to 20");
15             Screenshot("pcfcomponent_end.png");
16
17   testSettings:
18     locale: "en-US"
19     recordVideo: true
20     headless: false
21     browserConfigurations:
22       - browser: Chromium
23
24   environmentVariables:
25     users:
26       - personaName: User1
27         emailKey: user1Email
28         passwordKey: user1Password
```

Figure 9.16 – Test case steps for PCF component testing

- In the `testSettings` section, since the headless parameter defaults to `true`, we will adjust it to `false`. This tweak allows us to monitor browser interactions during test execution.

- The remaining parameters, such as recording video and the browser, can be left as-is. As for authentication, it should remain unchanged, but ensure the names of the environment variables match those you're using.

Running the test suite

We're now prepared to execute our test suite. Navigate to the folder where you downloaded the Test Engine GitHub repository. From there, locate the `/src/PowerAppsTestEngine` folder and execute our standard command:

```
dotnet run
```

If everything runs as expected, we'll observe the value changes in our browser and receive a final summary of the test results in our command line, as shown in the following figure:

```
Windows PowerShell            ×    +  ∨                                                    —    □    ×

Windows PowerShell
Copyright (C) Microsoft Corporation. All rights reserved.

Install the latest PowerShell for new features and improvements! https://aka.ms/PSWindows

PS C:\te3\PowerApps-TestEngine\src\PowerAppsTestEngine> dotnet run
Unable to parse log level: , using default
Running test suite: PCF Component
    Test results will be stored in: D:\results\pcfcomponent\5364e7
    Browser: Chromium
    App URL: https://apps.powerapps.com/play/e/8febb20b-0bb2-eba8-ae06-c5790023ca49/an/new_pcfcomponent_bcc03?tenantId=be
66fb28-a88d-47ff-8838-1f33b8d0bbbf
Test case: Case1
    Result: Passed

Test suite summary
Total cases: 1
Cases passed: 1
Cases failed: 0
Test results can be found here: D:\results\pcfcomponent\5364e7\Results_5364e7c6-9a2f-4c6e-9d4f-fbb35e7070f0.trx
PS C:\te3\PowerApps-TestEngine\src\PowerAppsTestEngine>
```

Figure 9.17 – Test execution summary for the PCF component

Checking the results

Just as we did with the canvas component testing, we can inspect the results at various levels: for the test suite, we can do this with an included video, while for the test case, we can do this with included screenshots. This can be seen in the following figure:

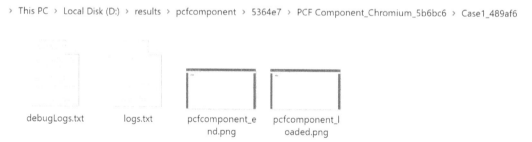

Figure 9.18 – The folder structure for the results of executing the PCF component

Even though the supplied sample is not complex, it offers a glimpse into how we can automate testing for PCF components included in canvas apps.

This can be paired with standard interactions with other controls so that we can comprehensively test end-to-end processes. This is particularly beneficial given that we can also mock network connections and telemetry measurements, as we'll explore in subsequent chapters.

Summary

This chapter provided an overview of *testing canvas components and PCF components* with Test Engine. First, we covered the basics of Canvas components, walking through an example of creating a *reusable UserCard component*. Then, we demonstrated how to test this component using Test Engine by modifying its properties, such as status and user role in a YAML test file, executing the test, and reviewing the results.

Then, we explained PCF components, which allow developers to create custom controls that extend the capabilities of Power Apps. We showed you *how to add a PCF component to a canvas app* and configure its properties.

To test the PCF component, we used a YAML test file to alter properties, ran the test suite with Test Engine, and inspected the logs, videos, and screenshots produced.

Overall, you learned how Test Engine enables automated testing of canvas and PCF components by allowing you to interact with their properties. This is advantageous compared to Test Studio, which cannot test these components. Through the practical examples provided, you learned how to create custom component tests and execute them with Test Engine to validate Power Apps that contain these components.

In the forthcoming chapter, we will broaden the scope of automation by covering essential concepts surrounding *testing automation with Azure DevOps pipelines*. Additionally, we will explore how to leverage these pipelines to automate tests constructed in Test Studio.

10

ALM with Test Studio and Test Engine

Given that our focus is on testing automation, it's important to delineate its relationship with DevOps processes. Incorporating *automated testing in Azure DevOps pipelines* offers several advantages, such as a faster testing process, continuous testing, and streamlined build and deployment. This integration enables the *continuous testing, building, and deployment* of iterative code changes, helping to catch failures ahead of production and improving overall software quality and productivity.

In this chapter, we will explore how to harness the power of Azure Pipelines to automate the testing of Power Apps. We will see how it can be leveraged to automate the testing of Power Apps applications built using the Microsoft Power Apps Test Studio test that we built in previous chapters.

First, we will set up the infrastructure required for test automation, including GitHub repositories and Azure DevOps. Then, we will create test automation pipelines using both the classic editor and YAML syntax. These pipelines will execute Power Apps test cases and suites stored in Test Studio and publish meaningful test reports.

We will also learn how to configure triggers to execute automated tests on code changes and scheduled intervals. Detailed test reports can provide insights into application quality and detect regressions. Finally, we will discuss the future evolution of the **Power Platform CLI tool** to simplify Power Apps testing.

By the end of this chapter, you will know how to implement end-to-end automation of Power Apps testing using Azure Pipelines. The key topics include the following:

- **Creating a classic editor and YAML pipelines for Power Apps test automation**: How to create Azure DevOps pipelines that include test automation parts

- **Configuring pipeline triggers for continuous testing**: Automating the triggering process of Azure DevOps pipelines, based on continuous integration or scheduling

- **Analyzing test reports and execution runs**: How to inspect the test results within the pipeline report and review artifacts such as screenshots

- **Understanding how the next evolution of Power Platform CLI can optimize test automation**: How the new release of PAC CLI, which includes the `tests` command, could impact automated testing processes

With these skills, you will be empowered to leverage Azure Pipelines to automate testing, scale test executions, and gain valuable insights into your Power Apps applications. Let's get started!

Technical requirements

In this chapter, we're going to utilize the Power Apps Test Studio tests that we covered in previous chapters, along with the associated requirements, such as the Power Platform environment and the sample application that we built in *Chapter 5* (`https://github.com/PacktPublishing/Automate-Testing-for-Power-Apps/tree/main/chapter-05`).

In addition to these, we will work with Azure DevOps pipelines. Therefore, you'll need to set up an Azure DevOps environment (if you haven't done so already) and a GitHub account so that you can fork the repository provided by Microsoft or the one we're providing for this book with some fixes.

We won't delve into the details of how to create these elements, which you can create for free, but we'll give you a brief introduction along with additional resources for further information.

Creating a GitHub account

Creating a **GitHub account** is a simple process that enables you to collaborate on software development projects and access a wide range of features. We will utilize GitHub to leverage the code repository that can be incorporated as a source in our pipeline, enabling the execution of test definitions created in Test Studio. We will describe this process in the subsequent sections.

To create a GitHub account, visit `https://github.com/join` and provide a username, email address, and password. After signing up, you'll receive an email to verify your account. Once verified, with the free account, you can enjoy unlimited access to public repositories and collaborate with up to three users on private repositories.

Creating an Azure DevOps environment

Azure DevOps is a key element for testing automation pipelines as it offers an integrated environment for version control, continuous integration, and deployment, streamlining the testing process.

Setting up an **Azure DevOps environment** is also a straightforward task. The best part is that you can get started for free, and there are no additional requirements for this book. To set up an Azure DevOps environment, you'll need to sign up with a Microsoft or GitHub account and then create a new project within your organization. You can choose the visibility level (public or private) for your project and invite team members to collaborate. Azure DevOps offers a wide range of features, including source control, work tracking, continuous integration, and continuous delivery, both on-premises and in the cloud.

To sign up, visit `https://azure.microsoft.com/services/devops/`. For further details on how to set it up, visit `https://learn.microsoft.com/en-us/azure/devops/user-guide/sign-up-invite-teammates?view=azure-devops`.

Exploring key concepts of test automation within Azure DevOps pipelines

As emphasized in previous chapters, test automation is a pivotal aspect of the software development process. When integrated with DevOps pipelines, it offers a host of additional benefits that enhance the overall efficiency, reliability, and quality of the software development process.

Before we delve into the specifics of executing automated tests within DevOps pipelines for Power Apps, let's brush up on some key concepts that we'll be discussing in the following sections.

Azure Pipelines

Azure Pipelines is a cloud-based service that facilitates the building, testing, and deployment of your applications. It provides a plethora of features, including continuous integration, continuous delivery, and support for a variety of programming languages and platforms. For instance, a trigger (such as a daily schedule) initiates the pipeline, which then progresses through various stages, each containing jobs to be executed by agents. These jobs are broken down into steps and tasks, which could range from building the code to running tests, and eventually yielding artifacts, the final products of the pipeline.

Test automation in Azure DevOps

Azure DevOps furnishes tools and services for automating test cases in your test plans. You can run these directly from **Azure Test Plans** (which requires a license) or within the pipelines you create, offering a robust orchestration workflow to obtain test binaries as artifacts and execute tests as part of the CI/CD process.

The PowerAppsTestAutomation project on GitHub

The **PowerAppsTestAutomation** project is a Microsoft initiative on GitHub aimed at providing an abstraction of Selenium tests. This helps automate operations such as signing into your application, launching a browser on the build agent, executing test cases and suites from Test Studio, and viewing the status of test execution in the Azure DevOps pipeline.

For this book, we'll be using our own fork of this project to address some issues that have arisen with the new Power Apps home page. We will delve into the specifics of these topics in the upcoming sections. Next, we'll examine the process of incorporating Power Apps tests, created with Test Studio, into an Azure DevOps pipeline.

Automating tests built in Test Studio with Azure Pipelines

In this section, we will cover the necessary steps to enable the Power Apps Test Studio pipeline in an Azure DevOps environment using both the classic editor and the new YAML approach. We will assume that you have already set up an Azure DevOps environment, as mentioned in the previous section, and have tests prepared in Test Studio, as covered in *Chapter 6* of this book.

We will begin with the common steps required for both the classic and YAML pipelines, then move on to the specific steps for each.

Before you start, it is necessary to complete the following steps – we'll delve into these steps in detail later on:

1. Fork the `Microsoft/PowerAppsTestAutomation` project on GitHub (`https://github.com/microsoft/PowerAppsTestAutomation`). As we will discuss later, there have been changes to the Power Apps home page that may cause errors when using the original project. As we are creating a fork of this open source project, you can implement these changes in your own fork. Alternatively, for convenience, you can directly fork a repository that already includes these changes for this book: `https://github.com/CesarCalvoCobo/PowerAppsTestAutomation`. We will explain the required changes anyway as they are not overly complex.

2. Create a new `TestURLs.json` file in the repository, listing the applications test URLs you wish to run from the pipeline. We will explain how to get these URLs in the next section.

Now, let's dive into the details of these steps.

Forking a repository

Forking a GitHub repository involves creating a duplicate version of a full repository belonging to another user. By forking, the entire contents of the original repository are replicated into your own GitHub account, as if you now possess your very own instance of that project. This copying of a repository from one account to another mirrors the familiar computing process of duplicating a folder from one location on a drive to another destination drive. Through forking, the projects of others can serve as the foundation for your repositories.

Forking allows you to freely experiment with changes without affecting the original project, which is a common practice in open source projects to enable contributions from any developer without having to manually manage authorization.

To fork a repository, you should follow these steps:

1. Navigate to the chosen repository on GitHub.

2. In the top-right corner of the page, click the **Fork** button.

3. Select your GitHub account as the owner of the forked repository.

4. The forked repository will be created in your GitHub account with the same name as the original repository.

Once you have forked the repository, you can clone it to your local machine and make changes to the code. After making changes, you can push them to your forked repository and create a pull request to propose your changes to the original repository. *Figure 10.1* shows a screenshot of our repository fork:

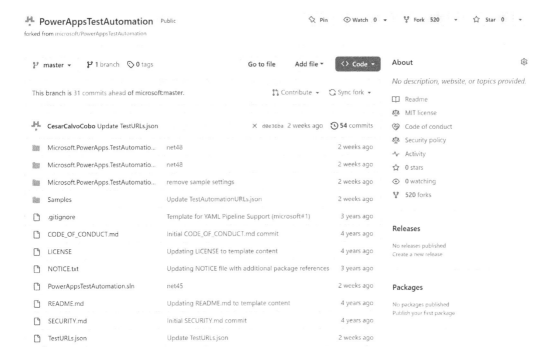

Figure 10.1 – Fork of the PowerAppsTestAutomation project

Let's delve into the specific process of forking our project, **PowerAppsTestAutomation**.

Forking the PowerAppsTestAutomation project

A fork is a copy of a repository. By forking the PowerAppsTestAutomation project, you can make changes without affecting the original project. Follow these steps to fork this project:

1. Go to the Microsoft/PowerAppsTestAutomation (`https://github.com/microsoft/PowerAppsTestAutomation`) repository on GitHub or the one that we created for this book with the fixes already included (`https://github.com/CesarCalvoCobo/PowerAppsTestAutomation`).

2. Locate and click the **Fork** option situated at the top-right corner of the repository's page.

3. Choose your personal GitHub account as the destination to duplicate the forked repository. Give it a name personalized for your repository. This will create a distinct copy separate from the original.

In our fork, we made some changes to make the PowerAppsTestAutomation project work. As the Power Apps home page has recently changed, it seems to be breaking some of the **xpath selectors** used in the original project. The two that we have changed to make it work are as follows:

*   In `RunTestAutomation.cs` and `OnLineLogin.cs.Reference.Login.MainPage`, which was using `//div[contains(@class,"home-page-component")]`. We switched it to `//div[@id='ppuxOfficeHeaderSearchBox_container']`

*   In `RunTestAutomation.cs` and `OnLineLogin.cs`: Regarding `By.Class Name("apps-list")`, we updated it by creating a new property in `Element Reference.cs` called `Login_AppsList` that points to `//div[@aria-label= 'Your apps']`.

In our fork, we have also updated the framework to .NET 4.8 to resolve various issues associated with the classic pipeline mode. We have also updated some NuGet packages, such as Selenium WebDriver and Selenium ChromeDriver. However, it's important to note that our main focus was to ensure compatibility with the Selenium Chrome drivers.

Again, for your convenience, you can directly use the fork that we have created, but check the updated information at the time you're reading this book.

> **Note**
>
> Please review the most recent information in the original Microsoft repository since potential solutions to these issues may already exist. Be aware that additional fixes might be necessary if the Power Apps interface changes after the publication of this book. Any adaptations needed based on these changes should be made at your discretion and without holding the authors of this book responsible.
>
> At the time of your reading, you may be able to utilize the new `pac CLI` in your pipelines, which now includes Test Engine's capabilities. Therefore, ensure to keep checking the latest information.

Creating a TestURLs.json file

The `TestURLs.json` file is a configuration file that's used in the Power Apps Test Studio pipeline to specify the URLs of the Power Apps applications you want to test. This file is necessary for the pipeline to know which applications to run the tests against. The file should be created in the forked PowerAppsTestAutomation repository and should contain the App Test URLs in JSON format.

Create a new `TestURLs.json` file in the forked repository with the App Test URLs you want to run from the pipeline. This file will be used to configure the pipeline to run the tests for your specific Power Apps application.

> **Note**
>
> You can find a sample of this file in the `Samples` folder of the original repository or your fork (`https://github.com/microsoft/PowerAppsTestAutomation/blob/master/Samples/TestAutomationURLs.json`).

To acquire the URLs to be used in this file, you will need to navigate to Test Studio and access the tests that we created in previous chapters. Here is the process:

1. In Power Apps Test Studio, select either a test suite or a test case.
2. Choose the **Copy play link** option.
3. If there are any unpublished modifications, you will be prompted to publish your tests.

After copying, the play link will be ready for use. This is the link that you should use in our `TestURLs.json` file. *Figure 10.2* shows the `TestURLS.json` files with two links that were obtained this way:

Figure 10.2 – TestURLs.json with two test play links

The finalized `TestURLs.json` file should be situated in the root folder of your repository, as depicted in the preceding screenshot.

If you're utilizing a fork of the repository provided in this book, you may be able to edit the file already residing in the root folder. However, if you're employing a fork of the original repository, it will be necessary for you to upload this file. This upload can be accomplished through the web interface, GitHub desktop, or a Git client. Regardless of the path you choose, it's crucial to commit the changes of this file to the master branch.

With the links to our tests now prepared, let's focus on the actual Azure pipeline.

Automating tests built in Test Studio with Azure Pipelines using the classic editor

In this section, we will delve deeper into the process of setting up Azure Pipelines with GitHub authentication and employing the *classic pipeline* variant. This procedure entails creating a new pipeline in Azure DevOps, linking it to your forked GitHub repository, and establishing the required authentication.

Begin by signing into your Azure DevOps organization, created as per the steps described earlier. If you have not done so yet, you may opt to create a new project; navigate to the project and select **Pipelines** from the left-hand menu. On the screen titled **Where is your code**, choose the **Use the classic editor to create a pipeline without YAML** option located at the bottom of the page. From here, select **GitHub** as the source of your code.

> **Note**
>
> Please verify your organizational settings to confirm whether the **Classic Pipelines** feature is enabled. Note that it is now possible to disable this feature at an organizational level through the settings (`https://devblogs.microsoft.com/devops/disable-creation-of-classic-pipelines/#enable-the-feature`).

At this juncture, you will need to authenticate with GitHub to permit Azure Pipelines access to your forked PowerAppsTestAutomation repository. Follow these steps to authenticate:

1. If you are not already signed into your GitHub account, you will be prompted to do so.

2. After signing in, you may be asked to authorize Azure Pipelines to access your GitHub repositories. Click **Authorize** to grant the necessary permissions.

   Upon authentication, you are now able to select the forked **PowerAppsTestAutomation** repository from your GitHub account, as shown in the following screenshot:

Figure 10.3 – Selecting the GitHub repository in our pipeline

3. Next, you will need to choose the template to be used for this pipeline. For this example, you may select the **Empty pipeline** template.

By adhering to these steps, you will successfully create a new Azure pipeline connected to your forked GitHub repository and authenticate with GitHub to grant Azure Pipelines access to your repository. Once the pipeline has been created, you can move forward with configuring the pipeline to use the `TestURLs.json` file and adding the necessary tasks for test automation.

In the subsequent step, we will need to choose the template for the pipeline. We have the option to select **Empty Pipeline** so that we can add our own steps.

Now, let's start creating the required steps in our pipeline.

When we create our pipeline, we will observe that *a default agent job has been created*. In the following sections, we'll utilize this default agent job.

In the agent pool, we'll retain the default value that is inherited from the pipeline. This value is associated with the agent pool and agent specification that we'll find in the pipeline we created by default: Azure Pipelines and Windows latest.

Azure DevOps distinguishes between two main types of agent pools: **Azure Pipelines agents** and **Private agents**.

Before delving into the tasks that should be included in the pipeline, let's first clarify the two types of agents involved:

- **Azure Pipelines agents**: These are Microsoft-hosted agents that provide a practical solution for executing your jobs. These agents simplify the maintenance and upgrade process as Microsoft handles them for you. Each time you run a pipeline, a fresh virtual machine is allocated for each job within the pipeline. After completing a job, the virtual machine is discarded, meaning any modifications made to the filesystem during a job won't be accessible to subsequent jobs.

> **Note**
>
> The free tier includes one Microsoft-hosted job with 1,800 minutes per month at the time of authoring this book. However, please verify the latest updates and any other requirements via the Microsoft documentation (`https://azure.microsoft.com/en-us/pricing/details/devops/azure-devops-services/`).

- **Private agents**: On the other hand, private agents are **self-hosted agents** that you install and manage on your infrastructure, such as on-premises machines or cloud virtual machines. These agents are useful when you require more control over the environment and access to local infrastructure or specific software.

Private agents can be particularly beneficial for the automation of UI tests. You can watch the various actions in the browser while executing the tests and have easier access and modification abilities of the test library. However, for the example in this book, it would necessitate more preparation. Therefore, for simplicity, we'll stick to Azure Pipelines agents, as shown in the following screenshot:

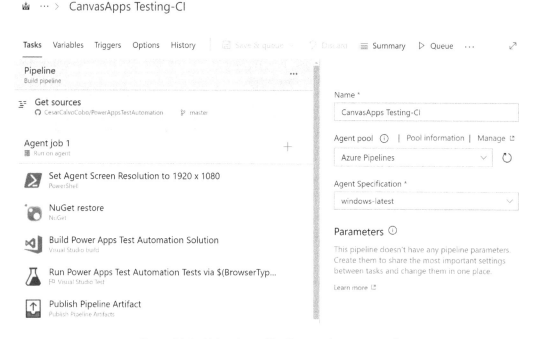

Figure 10.4 – Using Azure Pipelines as the agent pool

By choosing **windows-latest** as the agent specification, we opt for the most recent Windows version, which at the time of writing is Windows Server 2022. This choice ensures compatibility with the latest software, tools, and technologies. We could also use Windows 2019 for this example. However, sticking to the latest version is recommended to cater to potential future needs.

Now, let's begin with the tasks that we will incorporate into the pipeline.

Task 1 – configuring screen resolution using PowerShell

To circumvent issues when identifying certain DOM elements (since the PowerAppsTestAutomation project utilizes Selenium), one way to prevent errors when locating elements is to use a conventional resolution. Consequently, we will incorporate a PowerShell task to manage the logic for this in our Agent Job:

1. Add a PowerShell task to the pipeline.

2. Set the display name to `Configure Screen Resolution`.

3. In the **Script** field, enter the code shown in the following screenshot, which is also available in the Microsoft documentation (`https://learn.microsoft.com/en-us/power-apps/ maker/canvas-apps/test-studio-classic-pipeline-editor#step-1-- -configure-screen-resolution-using-powershell`):

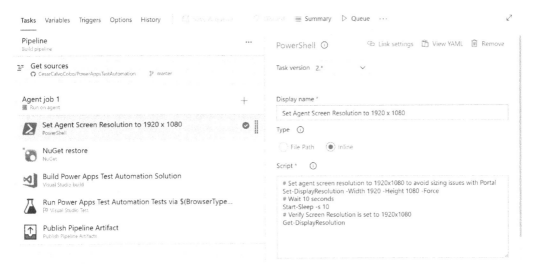

Figure 10.5 – Setting the screen resolution

Now, let's continue with our next task to get the NuGet packages that we need.

Task 2 – NuGet

An essential step in the pipeline involves enabling the **NuGet** packages necessary for compiling the **PowerAppsTestAutomation** solution. Restoring NuGet packages in an Azure DevOps pipeline entails downloading and installing the required dependencies for a project during the build process. These dependencies, which are specified in the project file or a `packages.config` file, need to be restored to ensure that the build and test environments possess the necessary libraries and tools to function correctly.

In our case, the NuGet packages include, for instance, the Selenium packages that enable web UI automation. Let's add this:

1. Add a **NuGet** task to the pipeline.

2. Set the display name to `Restore NuGet Packages` or a similar meaningful name.

3. Configure the task to restore the NuGet packages by selecting the path of the PowerApps TestAutomation solution:

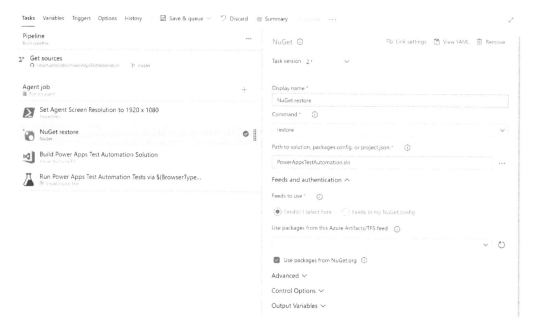

Figure 10.6 – Restoring NuGet packages

Once we have obtained the NuGet packages, we are ready to start building our solution.

Task 3 – building the PowerAppsTestAutomation solution task

We will use the **Visual Studio Build task** to build the **PowerAppsTestAutomation** solution. The aim of the Visual Studio Build task in an Azure DevOps pipeline is to construct your project using **MSBuild** and set the Visual Studio version property. This task automates your application's build process, ensuring that it compiles correctly and is ready for deployment or further testing:

1. In the pipeline editor, click on the + icon to add a new task.

2. Add a **Visual Studio Build** task to the pipeline.

3. Set the display name to `Build PowerAppsTestAutomation Solution`.

4. Configure it to use Visual Studio 2022. The Visual Studio version in the Visual Studio Build task in an Azure DevOps pipeline refers to the version of Visual Studio used to construct your project. Visual Studio 2022 is the latest version and works well with our fork, but Visual Studio 2019 can be another valid option, especially if you encounter any issues when working with .NET 4.5 in your fork.

5. Configure the task to build the PowerAppsTestAutomation solution:

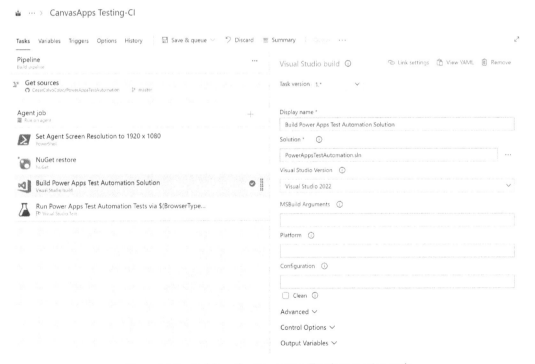

Figure 10.7 – Building the PowerAppsTestAutomation task

We are now prepared to configure our final task in the pipeline, where we will set up the actual testing component.

Task 4 – adding Visual Studio tests

The **Visual Studio Test task** in an Azure DevOps pipeline is used to run unit and functional tests, such as **Selenium** and others, using the **Visual Studio Test (VSTest)** runner. Let's examine the steps necessary to configure this component:

1. In the pipeline editor, click on the + icon to add a new task. Search for `Visual Studio Test` in the task search box and select **Visual Studio Test** from the list.

2. Set the display name to `Run Tests on Google Chrome` or something similar.

3. Regarding the choice of test selection, several options are available – that is, **Test assembly**, **Test plan**, and **Test run**:

   - **Test assembly**: This configurable setting indicates the test assembly files to be executed, which contain your assembled test code. Filters can also be provided to only run specific subsets of tests within the assemblies.

   - **Test plan**: By selecting this option, any automated tests connected to test plans will be invoked and run. This leverages your existing test plan configuration.

   - **Test run**: This mode is intended for ad hoc test execution outside of a CI/CD pipeline. It prepares the necessary test environments based on a chosen test plan. For automated testing within continuous integration or deployment, the test assembly or test plan options are recommended over the test run mode.

   We will use the **Test assembly** option, selecting our compiled assembly from the **PowerApps TestAutomation** project.

4. To select this assembly, we will point to the DLL that will act as the output path of the project compilation. In this case, it will be `**\Microsoft.PowerApps.TestAutomation.Tests\bin\Debug\Microsoft.PowerApps.TestAutomation.Tests.dll`.

5. We will look for it in `$(System.DefaultWorkingDirectory)`, a predefined variable in Azure DevOps pipelines representing the local path on the agent where your source code files are downloaded, and where the compilation will be produced.

6. We need to choose a folder where we will store our test results. The default value is `$(Agent.TempDirectory)\TestResults`. `$(Agent.TempDirectory)` is a predefined variable in Azure DevOps pipelines representing a temporary folder on the agent machine. This folder is cleaned up after each pipeline job, making it a suitable location for storing temporary files or artifacts during the pipeline's execution.

7. We will establish a filter for the tests to be executed. In our case, it will be `TestCategory=PowerAppsTestAutomation`.

8. Finally, we will choose the **Test mix contains UI tests** option to enable the agent to run in interactive mode:

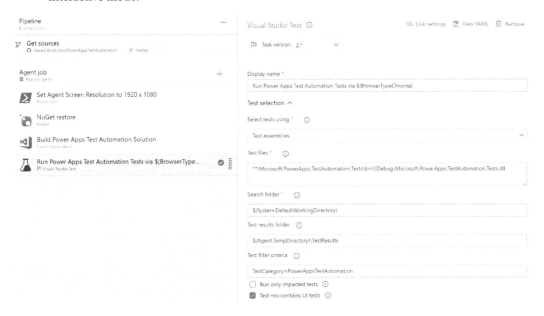

Figure 10.8 – Run tests

9. We will set our test platform version to **Latest**, which is **Visual Studio 2022**, but you can choose another option if you are not using Visual Studio 2022 as the VS version.

We need to choose the `testsettings` file. The `testsettings` file in an Azure DevOps pipeline test task is a configuration file that allows you to customize various aspects of your test execution, such as test run parameters, data collectors, and test adapters. You can override certain settings with variables to customize the test execution environment; we will do this later when we set our pipeline variables.

The `testsettings` file, which crucial to our example, resides at the following path, which must be specified when you're setting up `runteststasks`:

`Microsoft.PowerApps.TestAutomation.Tests/patestautomation.runsettings`

This holds true for both the original repository, as well as the modified version featured in this book. For guidance on configuring this step, refer to the following screenshot:

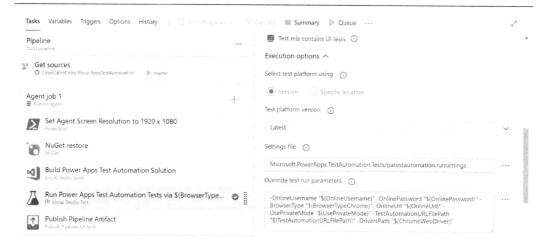

Figure 10.9 – Test settings file

10. Lastly, we will set our parameters that will override the testsettings file with our pipeline variables. We could avoid this step if we directly set the values in the testsettings file, but then the values would be stored in the GitHub repository. This way, the values can reside in our pipeline and can be configured differently for different pipelines.

The way of overriding these parameters is present in *Figure 10.9* and corresponds with this code:

```
-OnlineUsername "$(OnlineUsername)" -OnlinePassword
"$(OnlinePassword)" -BrowserType
"$(BrowserTypeChrome)" -OnlineUrl "$(OnlineUrl)" -
UsePrivateMode "$(UsePrivateMode)" -
TestAutomationURLFilePath
"$(TestAutomationURLFilePath)" -DriversPath
"$(ChromeWebDriver)"
```

The values with the $ symbols are pipeline variables that we need to configure in the next section. Let's explore how to configure these.

Task 5 – configuring pipeline variables

We will now configure the pipeline variables that were mentioned in the previous section, where we were overriding the testsettings file in the VSTest task we added earlier. To do this, follow these steps:

1. Navigate to the **Variables** tab.

2. Add the following variables:

    • Variable name: BrowserTypeChrome: Set its value to **Chrome**.

    • Variable name: OnlineUrl: This should point to the Power Apps home page. You can use either https://make.powerapps.com or https://create.powerapps.com.

- Variable name: `TestAutomationURLFilePath`: This needs to point to the `TestURLs.json` file that we created earlier. As per our fork, it is in the root folder of our sources directory: `$(Build.SourcesDirectory)\TestURLs.json`.

- Variable name: `UsePrivateMode`: Here, we can specify if we want the browser to navigate in private mode, which is often the best option for these types of tests where we might be switching users. Set the value to `true`.

- Variable names: `OnlineUsername` and `OnlinePassword`: Enter your credentials for the Power Platform environment. Remember to click on the lock to make these variables secrets.

Figure 10.10 shows an example of how to set these variables' values:

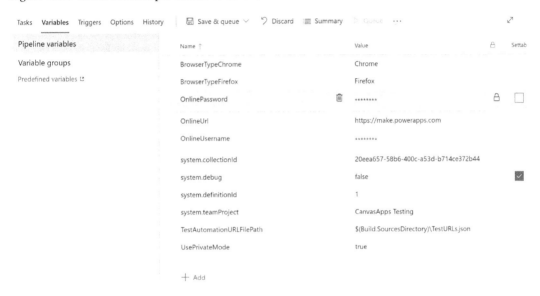

Figure 10.10 – Pipeline variables

We can save our pipeline by clicking on the dropdown labeled **Save & queue** and choosing the option titled **Save**. We will run this pipeline in the next section.

Task 6 – running the pipeline

Now, it's time to run our pipeline. We have two options for doing this. If we have made some changes, the **Save &Queue** option will be available in our editor. If we saved it previously, the **Queue** option will be enabled.

Selecting either of these will bring up a window where we can change our parameters to run the pipeline and finally trigger it to run the tests, as configured in the pipeline tasks.

Task 7 – monitoring the pipeline while it's running

You can monitor the progress of each task in real time. In the pipeline summary run, you'll see a list of jobs and tasks that are part of the pipeline. Each task will have a status indicator showing whether it's pending, in progress, completed successfully, or failed. Here's an example of a pipeline run:

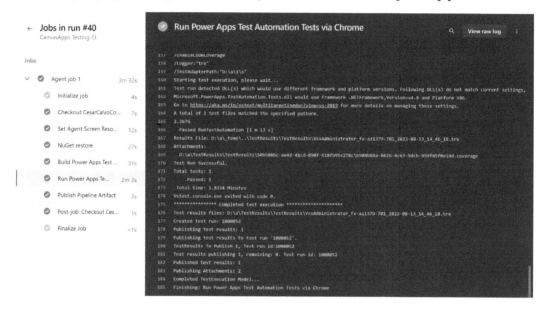

Figure 10.11 – Pipeline job run

Click on a specific task to view more details, such as the logs and any error messages. This can help you identify any issues or bottlenecks in the pipeline. By monitoring the pipeline while it's running, you can gain insights into the performance of each task and ensure that the pipeline is executing as expected.

Task 8 – monitoring test results

Once the tests have been executed, you can monitor the results and analyze the test execution in the **Tests** tab.

You can access the **Tests** tab by going back to the summary of the pipeline. This tab provides a detailed summary of the test's execution. Here, you will be able to check the percentage of tests passed, the run duration, and details of each test, with screenshots of every test after the execution attached to it.

> **Note**
>
> Please note that with this PowerAppsTestAutomation project, the test summary will report a single test result per browser. This test result will encompass one or more test cases or test suite results.

Most likely, your tests have passed, so you will need to filter the list to show the passed tests too (this particular filter can be located in the final column of the filter bar, situated beneath the **Column options** menu). However, you can filter here according to your preference, to show only failed tests, for instance. *Figure 10.12* illustrates the details of the test that passed in our pipeline with the information of every test case and screenshots attached:

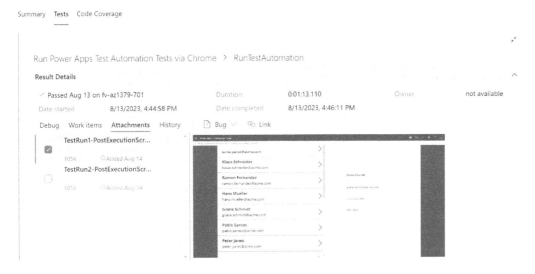

Figure 10.12 – Tests execution report

As we have discussed in this section, you can automatically check the status of your applications using Test Studio and Azure DevOps pipelines.

You can do this periodically or when you make changes to your code. You can enable these options in the **Triggers** tab when editing the pipeline, where you can enable continuous integration to be triggered when a commit is done in a certain branch, or you can schedule them (which might be more applicable when working with canvas apps and the PowerAppsTestAutomation project).

The following screenshot shows how to enable a scheduled trigger in the pipeline:

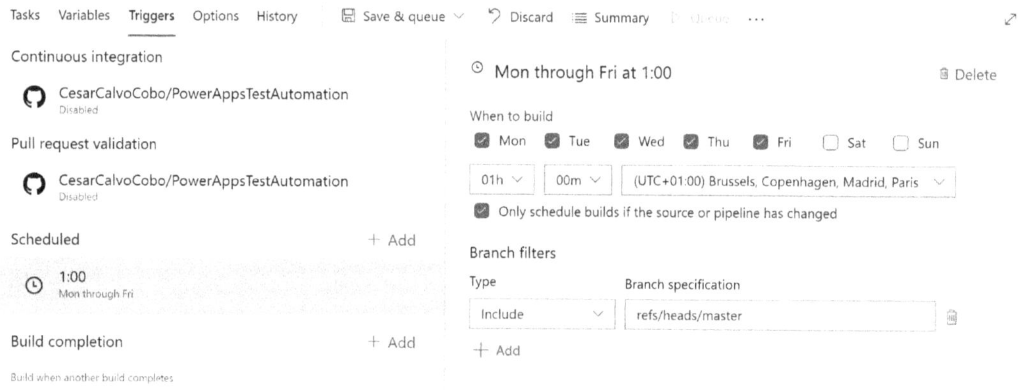

Figure 10.13 – Enabling a scheduled trigger in our pipeline

Having covered how to create a pipeline using the classic mode, let's delve into the process of creating it with a YAML pipeline, which seems to be the current focus with updates arriving earlier compared to the classic pipelines. However, the concepts will be very similar. We will use a YAML file to configure the tasks and the pipeline flow instead of a graphical interface.

Automating tests built in Test Studio with Azure Pipelines using YAML

Having already covered the steps for forking the repository and creating a `TestUrls.json` file (we will be using the same ones for this section), we will start by creating a new YAML pipeline:

1. Navigate to the **Pipelines** section and click on **New pipeline**.

2. Choose **GitHub YAML** as the source hosting platform and select your **PowerAppsTestAutomation** repository fork. You should already have authentication set up for GitHub, so you won't need to create a new one.

3. Under **Configure your pipeline**, select the **Existing Azure Pipelines YAML file** option and choose the `azure-pipelines.yml` file from your repository.

    In our fork, we have already created an `azure-pipelines.yml` file in the root folder, which we have copied and adapted from the one existing in the original repository – that is, `Samples/azure-pipelines.yml`.

Figure 10.14 shows how we select this YAML file:

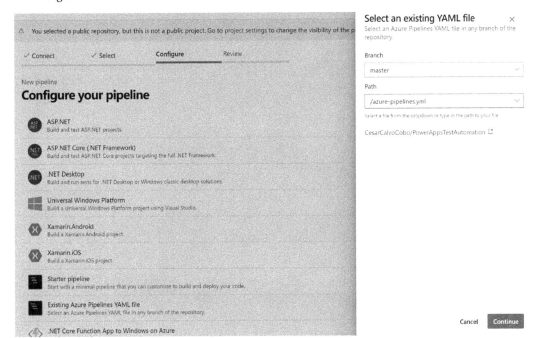

Figure 10.14 – Selecting the YAML file

4. You can choose this one or create your own. If you choose the one that we have created, you will need to update the repository name with *your own fork and the endpoint connection*. This is the name of the endpoint that you created to connect with GitHub from Azure DevOps.

Check out the example shown in the following figure:

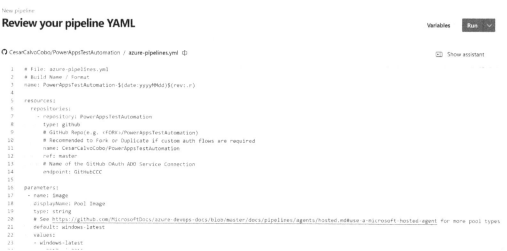

Figure 10.15 – Editing the YAML file

5.   Next, we will need to update several other values:

*   The portal URL, which is typically either `https://make.powerapps.com/` or `https://create.powerapps.com`.

*   The **Browser PrivateMode** setting, which we will configure to `true`.

*   **filename**, which is where the test URLs are stored in JSON format. In our scenario, this is `TestURLs.json`.

*   The **LocalProjectName** value, which should point to your repository name. In our case, it's `PowerAppsTestAutomation`, as we didn't change the name.

6.   If we click on variables, we can add both variables and keep them secret:

I.   Create a new variable called `OnlineUsername` and set its value to the Azure AD email address for the user account that will sign in during testing. Tests will execute using this user's context.

II.   Confirm the addition of the `OnlineUsername` variable.

III. Next, make another variable named `OnlinePassword` and set its value to be the password for the user account specified in the `OnlineUsername` variable.

IV.   This will allow the tests to authenticate and run under the provided credentials.

In the following screenshot, we are configuring one of these variables and the rest of the values described previously:

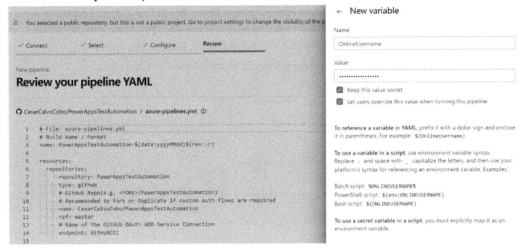

Figure 10.16 – Configuring variables and parameters for the YAML file

7. Next, we need to configure tasks in another YAML file. Within the YAML file, we will see another YAML file that has been mentioned (`frameworksteps.yml`) that will cover the steps that we created in our graphic interface, but this time in YAML format. The following screenshot shows the reference to `frameworksteps.yml`:

```
steps:
- checkout: self
- checkout: PowerAppsTestAutomation
#- template: localsteps.yml # Provide optional local YAML template here
- template: frameworksteps.yml@PowerAppsTestAutomation
  parameters:
    OnlineUsername: $(OnlineUsername)
    OnlinePassword: $(OnlinePassword)
    OnlineUrl: $(PortalUrl)
    BrowserType: ${{ parameters.BrowserType }}
    # File location format is /s/TeamProjectName/FilePathToTestAutomationURLs.json
    TestAutomationURLFilePath: $(Build.SourcesDirectory)\$(LocalProjectName)\$(TestUrlFileName)
    UsePrivateMode: $(PrivateMode)
    LoginMethod: ${{ parameters.LoginMethod }}
```

Figure 10.17 – The frameworksteps.yml file referenced

Here, we can observe the same tasks that we created in ok classic mode, but in our fork, we have removed the task covering the Firefox tests from the original file due to the issues mentioned previously in this chapter:

> **Note**
>
> For those utilizing our forked repository, both `azure-pipelines.yml` and `rameworksteps.yml` will already be configured. You can also find copies of these files in this chapter's repository. However, it's crucial to modify the repository name so that it matches your fork and adjust the endpoint to your authentication configuration within `azure-pipelines.yml`.
>
> Don't forget to create the pipeline variables pertinent to authentication as well.

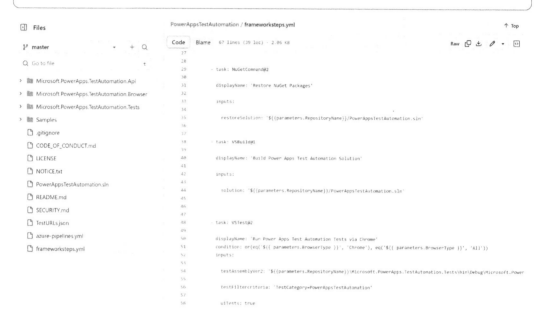

Figure 10.18 – frameworksteps.yml (tasks)

8. Returning to our pipeline creation, we will have the option to either save it or save it and run it directly. After clicking on **Run**, the pipeline will be saved and run automatically.

    Next, we will be able to view the summary of the execution run in Azure DevOps, with our agent immediately queued, as can be seen in the following screenshot:

Figure 10.19 – Pipeline execution summary

After initiating the pipeline, we can monitor the execution of tasks as well as the logs tracking our job run. These topics will be covered in the subsequent section.

Monitoring job run

Once we run the pipeline, we can monitor the results and analyze the test execution, the same as we did in the classic mode pipeline. We can go to the summary of the pipeline run and check the details of the agent job task:

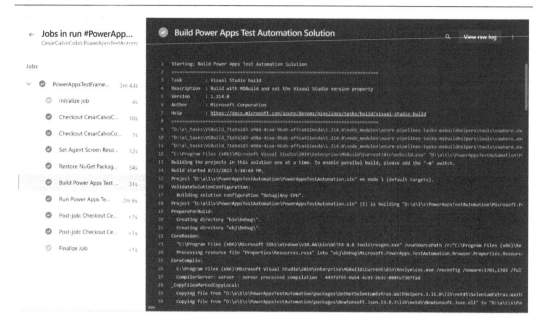

Figure 10.20 – Monitoring the job run

In this example, we have manually triggered the process. However, as previously mentioned, there are automated methods for running the pipelines. When a pipeline is created using a YAML file, the **Continuous Integration** trigger is enabled by default, but we can override this. We will explore this option in the forthcoming section.

Changing pipeline triggers

By default, the pipeline created with a YAML file will have the **Continuous Integration** trigger enabled. However, we can override this behavior and create a schedule trigger by editing the pipeline and accessing the **Triggers** section, as shown in the following screenshot. You can access this section by clicking on the three dots in the right column when editing the pipeline, and then clicking on **Triggers**:

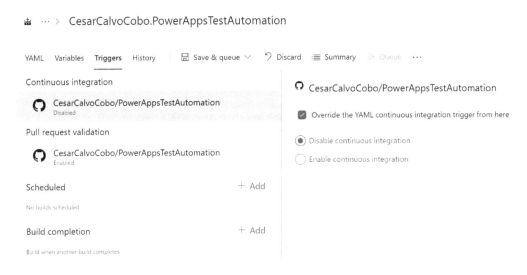

Figure 10.21 – Overriding continuous integration behavior in the YAML pipeline

Now that we've discussed the current process of automating Power Apps tests with Test Studio and Azure DevOps pipeline, let's explore potential future evolutions of this scenario. We'll specifically look at the new functionalities planned for **Power Platform CLI** and integrate Test Engine within it.

Evolution of Power Apps test automation pipelines with PAC CLI and Test Engine

Microsoft Power Platform CLI (PAC CLI) is a potent and comprehensive command-line interface designed to provide developers with the means to perform a wide array of operations within the Microsoft Power Platform ecosystem. This tool simplifies tasks related to environment life cycle management, authentication, and Microsoft Dataverse environments, among others. With Power Platform CLI, developers can automate tasks, tailor their development processes, and incorporate this tool into their existing workflows, thereby streamlining the development experience.

At the time of writing this book, Microsoft has rolled out a new version of PAC CLI (the current version being *1.26.5+g670cdf9*) that includes a preview feature that is highly relevant to this book.

This feature is the `pac tests run` command (refer to the Microsoft documentation at `https://learn.microsoft.com/en-us/power-platform/developer/cli/reference/tests#pac-tests-run`). It's built upon the **Test Engine** component that we've been exploring throughout this book, and it's poised to be the preferred method for executing Test Engine in the future.

As we underscored in *Chapter 5* and *Chapter 6*, this feature is still under preview but is highly likely to become the standard for conducting these tests. This is because PAC CLI is a comprehensive toolbox packed with numerous additional utilities.

More importantly, it is specifically pertinent to this chapter as it's projected to be the de facto tool to perform canvas apps automated tests within Azure DevOps pipelines. Despite its current preview status and the potential for changes, it's worth exploring the opportunities it could offer for continuous testing of canvas apps.

To acquire Power Platform CLI on your desktop, you can install it via the Power Platform Tools extension for Visual Studio Code, as a Windows MSI Installer, or as a .NET tool.

However, the primary focus of this chapter is the integration of Power Platform Tools with Azure DevOps. To achieve this, one method is to install **Power Platform Build Tools** in the Azure DevOps environment (you can find them at `https://marketplace.visualstudio.com/items?itemName=microsoft-IsvExpTools.PowerPlatform-BuildTools`). Once installed, you can point to the path of PAC CLI. Following this, you will be able to utilize the commands available within this tool. *Figure 10.22* illustrates the execution of PAC CLI without any additional command:

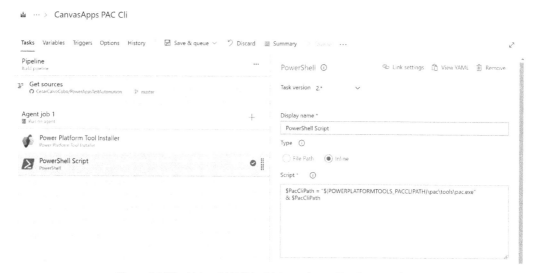

Figure 10.22 – Using PAC CLI within an Azure DevOps pipeline

Executing PAC CLI without any parameters will result in the tool displaying the list of available commands. *Figure 10.23* offers a glimpse of these commands as they stood at the time of writing this book, including the `tests` command in preview:

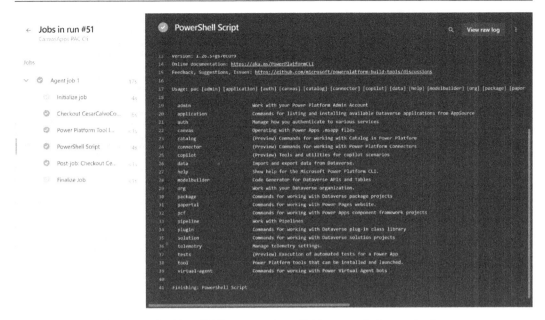

Figure 10.23 – List of PAC commands available currently in our pipeline

Microsoft has now *enhanced the usability of Test Engine by integrating it into the PAC CLI* toolset. Consequently, there is no longer a necessity to build the Test Engine GitHub project to conduct Power Apps tests. In addition, Microsoft is in the process of developing tasks for Azure DevOps Pipelines and GitHub Actions to support this mechanism.

At the time of writing, there's no existing documentation for employing this piece within Azure DevOps pipelines. Additionally, it seems to be currently non-operational for us – probably due to issues locating the correct executable paths of browser drivers. However, it could potentially be possible to execute our tests in the Azure DevOps pipeline by including the necessary parameters in the command, as shown in *Figure 10.24*:

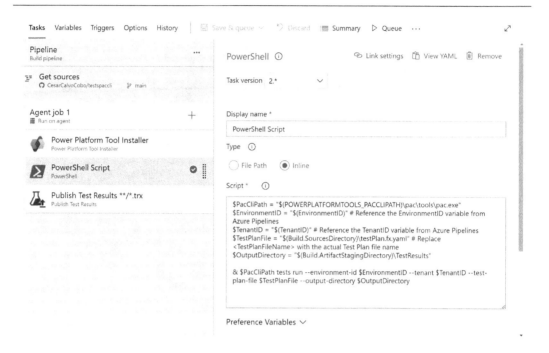

Figure 10.24 – PAC command to run tests

Nonetheless, as we mentioned earlier, it is anticipated that a specific task for Azure DevOps pipelines and GitHub actions, relevant to this functionality, will be released in the near future.

Consequently, *automated testing on Power Apps will be significantly simpler*. With Test Engine as part of Power Platform CLI, makers will be able to execute tests using these tasks within GitHub and Azure DevOps integrations.

It's important to keep an eye on the evolution of these components.

Summary

In this chapter, we explored how to leverage Azure DevOps pipelines to automate testing for Power Apps applications built with Microsoft Power Apps Test Studio.

We covered the steps to set up the required infrastructure, including GitHub repositories and an Azure DevOps environment. Next, we detailed the process of *creating Azure DevOps pipelines using both the classic editor and YAML syntax*. These pipelines execute Power Apps test cases and suites stored in Test Studio and publish meaningful test reports.

We also discussed *how to configure triggers* to run automated tests on code changes and scheduled intervals. Monitoring detailed test reports provides valuable insights into application quality and detects regressions.

Finally, we examined the *future evolution of Power Platform CLI tools to simplify Power Apps testing*. The integration of Test Engine into PAC CLI will streamline test execution with a single command. Azure DevOps tasks and GitHub Actions will also gain support for this test automation approach.

With these skills, you'll be able to leverage the capabilities of Azure Pipelines to automate testing, scale test executions, and gain critical insights into your Power Apps applications. In the upcoming chapters, we'll continue to delve into topics related to these automation processes, such as how to mock network connections in our tests and how to integrate telemetry from our application executions for deeper insights.

11

Mocks with Test Engine

In the previous chapter, we introduced you to Azure DevOps pipelines and their application in testing automation. We discussed some key aspects, such as running canvas app tests created in Test Studio using the Azure Pipelines classic editor and YAML. Furthermore, we learned how to incorporate tests that were built in Test Engine into Azure DevOps pipelines.

In this chapter, we will shift our focus to another critical aspect of testing – mock testing, specifically in Test Engine. In essence, mock testing allows test authors to create imitations of network calls, emulating the dependencies their application relies on. This practice is especially valuable when Power Apps makes calls to connectors, as it enables an app to be tested without any modifications and mitigates unwanted effects from external services.

The following sections will guide you through various aspects of mock testing:

- **What mock testing is in Test Engine**: You'll understand the necessity and application of mock testing, as well as its utilization in Test Engine

- **Mocking a SharePoint Service from Microsoft 365**: Learn how to execute a test where the app is dependent on a SharePoint Service

- **Mocking Dataverse in Your Power Apps**: Get equipped with the skills to run a test when an app relies on Dataverse

- **Mocking a Power Apps Connector**: Master how to perform a test where the app depends on a connector

By the end of this chapter, you will be well-versed in the concept and practice of mock testing in Test Engine. You will also understand how to carry out tests when your Power Apps have dependencies with SharePoint Service, Dataverse, and various connectors. Join us in this exciting journey to maximize the effectiveness of your Power Apps testing.

Technical requirements

The examples in this chapter are ordered based on their complexity and requirements, so the first example has fewer requirements and is easier to run, and complexity grows throughout the following examples. All examples can be run with minimal requirements, but to validate the full experience – that is, to check the running mock service and then the real service example – we recommend you try one with full requirements. To follow the examples in this chapter, you must download the corresponding files from the GitHub repository at `https://github.com/PacktPublishing/Automate-Testing-for-Power-Apps/tree/main/chapter-11` and follow its instructions. We will utilize three samples with the same app, available at the root of the repo, called `PowerParse_1_0_0_0.zip`; the examples are as follows:

- **SharePoint** and **Dataverse** services, used in the app, are provided for this chapter. Download the `testplans.zip` file from the repo, and use files from the `SharePoint` and `Dataverse` folders.

- And **HuggingFace AI Model BLOOM** service, configured in the app, to test generative AI examples. Download the `testplans.zip` file from the repo, and use files from the `HuggingFaceBLOOM` folder. AI Model will be invoked through a Power Automate flow, using the **Postman** Mock server API. Download the `testplans.zip` file from the repo and use files from the `Postman` folder.

Conversely, you will need the same **Power Platform environment** of previous chapters to install the examples provided, a text file editor such as **Visual Code**, and several accounts for the services to validate the mock routes we will discuss:

- To check the SharePoint example, you will need a working Microsoft 365 developer plan environment, as mentioned in previous chapters, from `https://developer.microsoft.com/en-us/microsoft-365/dev-program`.

- To test the generative AI examples, you can use a working Hugging Face free account: `https://huggingface.co`. Remember that this account is only required to run the actual results from the service when running the app, not for running the test.

- A Postman free account (`https://www.postman.com/`)to create a mock server API.

Due to the nature of the Test Engine, you will need the specific configuration of the environment. Let's take a look.

> **Note**
>
> The GitHub repository includes additional examples for generative AI, but in the book, only the minimum technical requirements will be presented. For example, although in the repository Azure OpenAI Service GPT API models (`https://azure.microsoft.com/en-us/products/cognitive-services/openai-service/`) will be included, in this chapter, you will use the **Hugging Face BLOOM API** (`https://huggingface.co/bigscience/bloom`) due to its free access API.

Finally, you will need to use Test Engine, so check *Chapter 6, Overview of Test Engine, Evolution, and Comparison*, and *Chapter 7, Working with Test Engine*, for details on the requirements and provisioning. We will mention a forked version of Test Engine for an advanced scenario. In this case, the only change is the executable used – namely, the one provided at `https://github.com/PacktPublishing/Automate-Testing-for-Power-Apps/blob/main/chapter-11/`, named `PowerAppsTestEngine_chapter_11.zip`.

Why utilize a forked version of Test Engine?

The existing version of Test Engine, particularly in the Mock scenario, presents certain limitations. To address this, we introduced an experimental version aimed at comprehending the various mock strategies in play. Review the platform as these limitations will be overcome, as well as some configuration needed later in the section.

Importing a solution

The **Power Parse PowerApps** solution is a sample app to run our network mock tests. In your Power Platform environment and with a working SharePoint site, to import the app, download it from the GitHub repository for this chapter, and follow these steps:

1. Navigate to `https://make.powerapps.com` and go to the **Solutions** page in the Power Platform developer environment. You can find this page by clicking on the **Solutions** tab in the left menu.

2. Click the **Import solution** button at the top of the page.

3. In the **Import a solution** window that appears, click the **Browse** button and select the ZIP file containing the solution (found in the GitHub repository – `PowerParse_1_0_0_2.zip`).

4. Continue clicking the **Next** button to begin the import process.

5. As we anticipated in the requirements, you will be asked for a SharePoint connection. Click **New Connection** and follow the wizard to create a new one. After creating the connection in a new window, come back to the import process to click **Refresh** and complete the solution import process.

6. The import process may take some time, and you will see at the top of the page a **Currently importing solution 'Power Parse'** message. Once the import is complete, the solution will appear in the list of solutions, and the SharePoint connection will be available in the list of connections, accessible through the left menu options.

In the following section, we will outline the process of importing some small datasets used by our `Power Parse` sample app, enabling us to execute test operations.

Configuring dependencies

To work with the application, there are some dependencies necessary to start using it. You can see those dependencies in *Figure 11.1* and its visual representation in *Figure 11.2*.

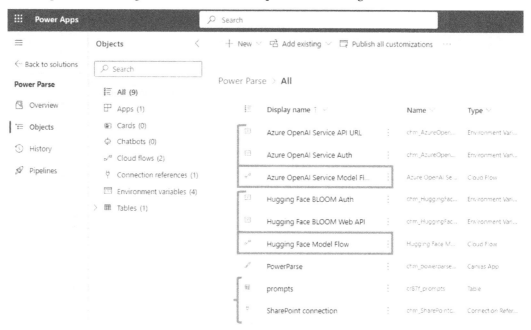

Figure 11.1 – Solution objects detail in the Power Parse app

The figure shows you the related solution artifacts details:

- **Dataverse** will be used to store prompts used as user instructions for the Azure OpenAI Service and Hugging Face services. The solution import process includes a `prompts` table.

- **SharePoint** will be used to store content used as context for the previous instructions. You will need to create a **list** with the `AI content` data.

- **Power Automate** Cloud Flows will be used to call AI services, and environment variables will hold generative AI services information, such as URL or API key.

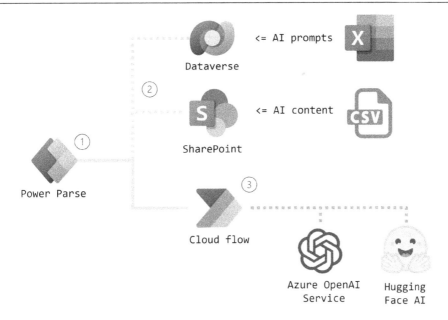

Figure 11.2 – Services and data dependencies in the Power Parse app

Demo data

Now, we need to address the configuration related to the information used in the application. In the **SharePoint** site, you created in the **Solution import** process, you will need to create a **SharePoint list** and import some sample content into it.

> **Note**
>
> A CSV file (`AI Content.csv`) containing four sample instructions is provided in the GitHub repository of this import process in SharePoint. However, the Power Parse app includes `add`, `delete`, and `edit` forms to manage all data.

The steps to complete the `AI content` data import process are as follows:

1. Sign in with your account at your **Microsoft 365** tenant at `https://<your-tenant>.sharepoint.com`, and click on **Create Site**. Select the team site, and add a name and description for it. Leave everything else as default. This site will store `AI Content`.

2. On the home page of your **AI Content** site, click on the **New** menu and then click **List**:

   I.   Click on **Create a list From CSV**.

   II.   Upload the `AI Content.csv` file from the GitHub repository.

   III. Click **Next** and then click on **Create**, leaving everything else as default.

The imported data should now be available in an **AI Content** list, as shown in the following figure:

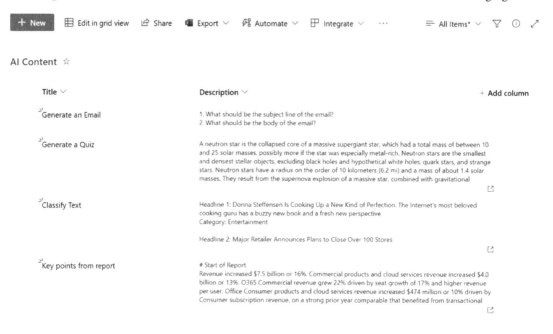

Figure 11.3 – Context for the AI instructions data imported

A **Dataverse** table was imported in the **Solution import** process, and we will need to populate sample content in it.

> **Note**
>
> An Excel file (`AI Prompts.xlsx`) containing four sample instructions is provided in the GitHub repository of this chapter. However, the Power Parse app includes `add`, `delete`, and `edit` forms to manage all data.

To add `AI prompts` data to the **Dataverse** table, click on the recent **Power Parse** solution and update the values of the `prompts` table mentioned in the previous steps:

1. Click on the `prompts` table, select the **Import** menu, and then the **Import data from Excel** option:

    I.   Click on **Upload**, and choose the `AI Prompts.xlsx` file from your computer.

    II. Click on the **Map columns** button to proceed to the mapping step.

    III. Assign the unmapped columns – map `description` to `description`, and `instruction` to `instruction`. After mapping, click on **Save Changes**.

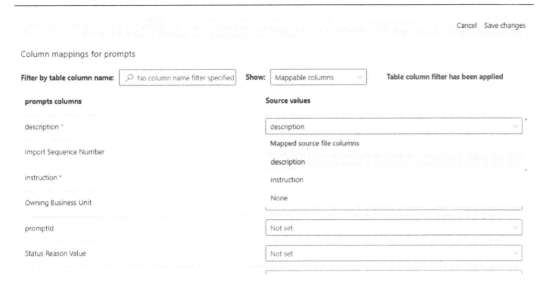

Figure 11.4 – Mapping demo data to the prompts table

IV. The mapping status message should say **Mapping was successful**.

V. Click on the **Import** button to start the import process.

VI. Wait for the import process to complete. A notification will be displayed, indicating the number of records imported successfully.

VII. Click on the **Close** button to finish the import process.

The imported data should now be available in the `prompts` table, as shown in *Figure 11.5*:

Power Parse > Tables > **prompts**

Table properties	⚙ 💼 ⌄	Schema ⓘ		Data experiences ⓘ		Customizations ⓘ
Name		▦ Columns		▤ Forms		⅗ Business rules
prompts		⚬ Relationships		▯ Views		▤ Commands
Type		🔍 Keys		⚋ Charts		
Standard				▦ Dashboards		

▦ prompts columns and data   ▦ Update forms and views   ✎ Edit | ⌄

🔤 description • ⌄	🔤 instruction • ↑ ⌄	🔤 promptId • ⌄	+14 more ⌄ +
Key points from report	Below is an extract from the annu...	5c115092-581e-ee11-9cbd-	
Classify Text	Classify the following news headli...	5b115092-581e-ee11-9cbd-	
Generate a Quiz	Generate a multiple choice quiz fr...	5d115092-581e-ee11-9cbd-	
Generate an Email	Write a product launch email for...	5e115092-581e-ee11-9cbd-	

Figure 11.5 – Context for the imported AI instructions data

Finally, remember two things – if you run the tests with another user, give read/write permissions to the list, and refresh the SharePoint connection in the **Connection references** section of the **Power Parse** solution, as shown in *Figure 11.6*. However, when you launch the app, it will request a valid connection.

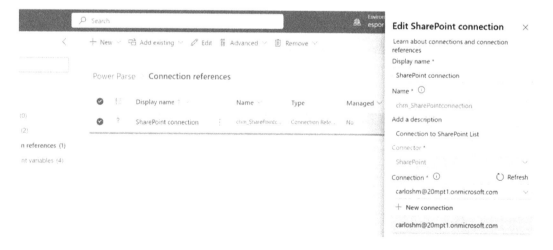

Figure 11.6 – The SharePoint connection edit form

Just take into account that the column title in the SharePoint list, should have the same value as the column description in the `Dataverse` table. All the content could have been set in one `datasource`, but we wanted to test both services, hence the need to link both on the title key for simplicity. Let's continue with the services configuration.

Services configuration

> **Important**
>
> API URLs and authentication information are stored as environment variables in the solution. Remember to edit them with information from your own environments if you want to use the app and get real data.
>
> You can use the app without **Azure OpenAI Service** or **Hugging Face** information, as long as you test the default and third mock strategy that does not use the actual service. However, running the actual app will fail due to not having the backend services.

To update AI service information, URL, and auth tokens, perform the following steps:

1. Navigate to `https://huggingface.co/bigscience/bloom` (you will need to sign up for a free account), and in the **Deploy** menu, click and select **Inference API**, This will bring up a popup with the information needed, as shown in *Figure 11.7*:

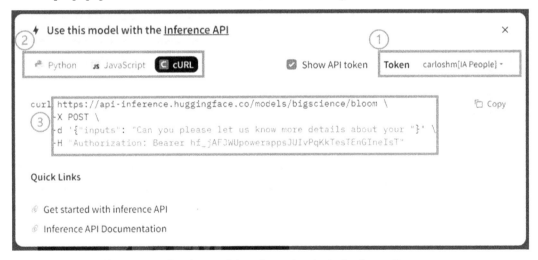

Figure 11.7 – Services and data dependencies in the Power Parse app

I.   Once you log in (**1**), you should click on **Show API token**.

II.   Then, select **cURL** tab to read the API information (**2**).

III.   Finally, **Hugging Face BLOOM Web API** will be `https://api-inference.huggingface.co/models/bigscience/bloom`, and **Hugging Face BLOOM Auth** will be the value after `Authorization` – in this example, `Bearer hf_jAFJWUpowerappsJUIvPqKkTesTEnGIneIsT` (3).

2.   To add this information to the solution, click on the recent **Power Parse** solution, and update the values of the environment variables mentioned in the previous steps:

I.   Click on each environment variable name, update the current value on the task pane on the right, and save it.

II.   The variables will be **Hugging Face BLOOM Auth** and **Hugging Face BLOOM Web API**, as shown in *Figure 11.8*.

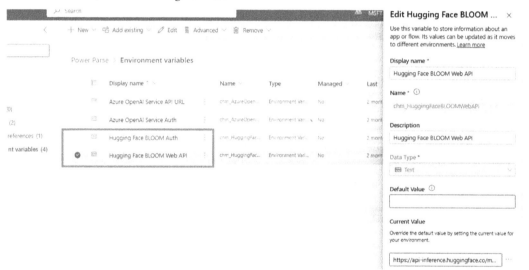

Figure 11.8 – The environment variables for generative AI services

The requirements could be a bit overwhelming, but you could start with the easier routes, explained in the chapter, and continue with the complex one as a bonus experience. Let's finish with an important note.

Power Platform Admin configuration

Test Engine is an experimental project that is being tested to incrementally be included in Power Platform to support testing capabilities. It has some limitations:

- It does not support multifactor authentication.
- You need to review controls used by the app, as there are several controls unsupported.

- It runs tests in a browser configuration that currently does not allow it to configure its state. For example, the connection consent dialog state is not preserved in tests `https://learn.microsoft.com/en-us/power-apps/maker/canvas-apps/connections-list#connection-consent-dialog`.

In the mock scenario, this third point is critical, so until it can be automated through **Test Engine**, you need to avoid the dialog. Connectors accessing a non-Microsoft, third-party service, do not allow to suppress the consent dialog. Luckily, we selected Microsoft first-party connectors that support single sign-on.

We will use `Set-AdminPowerAppsApisToBypassConsent` command (`https://learn.microsoft.com/en-us/powershell/module/microsoft.powerapps.administration.powershell/set-adminpowerappapistobypassconsent?view=pa-ps-latest`), the `environmentID` gathered in earlier or *Chapter 7*, and the `AppID` (`https://learn.microsoft.com/en-us/power-apps/maker/canvas-apps/get-sessionid#get-an-app-id`).

In an administrator **PowerShell** console, follow the next steps, and log in with your environment admin credentials:

```
PS > Install-Module -Name Microsoft.PowerApps.Administration.
PowerShell
PS > Add-PowerAppsAccount
PS > Set-AdminPowerAppApisToBypassConsent -EnvironmentName
environmentID -AppName AppID
```

This will allow our test to run without the consent dialog to appear.

Thus, with our environment and solutions prepared, we are ready to begin working with them. Let's proceed to the next section to start testing our application.

Getting to know mock testing in Test Engine

I will take a step back before diving into this topic to present the type of activities we can do in the preview version of **Test Engine**, available at the time of writing. Test Engine is based on **Playwright** (`https://playwright.dev/`), a library that provides cross-browser automation. It allows you to execute your tests, interact with elements, navigate to screens, or simulate API calls in your apps, as **Power Apps** runs on any browser as a web app. Test Engine uses Playwright to monitor and modify network traffic. Any request that Power Apps makes can be modified.

Let's imagine your app called GPT from Azure OpenAI Service to generate a summary of a text. You don't want to call the service every time in your tests, so in Test Engine, you will create a mock for the service. In *Figure 11.9*, you can follow what is going on under the hood:

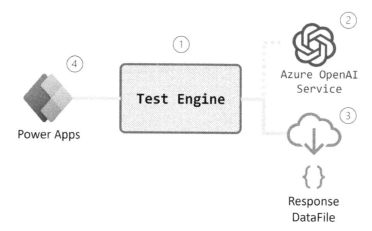

Figure 11.9 – The Test Engine flow to intercept network calls

1. First, as defined in the test file, the Test Engine runs the different steps to automate the execution of Power Apps in a selected browser.

2. For the aforementioned example, where the app calls the Azure OpenAI Service API to run the GPT model, the Test Engine will intercept the GPT call and make a different route.

3. Test Engine has a standard behavior of fulfilling a request. What does this mean? It means it will serve a static file with the response instead of calling the Azure OpenAI Service API.

4. Finally, the Power Apps app will get the response without noticing the change.

Let's describe this request-response process in the context of mocks.

HTTP request and response concepts

In the realm of web development and APIs, understanding the mechanics of HTTP requests and responses is crucial. Test Engine in our scenarios will intercept a **request**, check certain values, such as headers values and resource name, and change the response. In *Figure 11.10*, the client sends a GET request to retrieve a resource from `example.com`. It includes a custom header, **X-Custom-Request-Header**, with a `Value` value. The server responds with a `200 OK` status, indicating that the request was successful. The response includes a custom header, **X-Custom-Response-Header**, with a `ResponseValue` value.

GET /api/resource HTTP/1.1
Host: example.com
User-Agent: MyClient/1.0
Accept: application/json
X-Custom-Request-Header: Value

HTTP/1.1 200 OK
Date: Mon, 03 Oct 2023 00:00:00 GMT
Server: MyServer/2.0
Content-Type: application/json
Content-Length: 123
X-Custom-Response-Header: ResponseValue
{
  "key": "value",
  "message": "Resource successfully retrieved."
}

Figure 11.10 – A simple HTTP request-response flow

This foundational knowledge becomes even more significant when dealing with testing automation in Power Apps. Here, we will briefly delve into four key concepts that come into play while interacting with HTTP services:

- **Fulfilling the Request**: Successfully completing a client's HTTP request by processing it and returning the appropriate data or resource to the client

- **Continuation of the request but adding new headers**: Supplementing an initial request with additional headers before final processing

- **Rerouting the request**: Changing the request's destination from the defined service to another specified

- **Adding new headers**: When rerouting, adding new headers to provide additional information or instructions for the new destination server

Grasping these concepts will enable you to better simulate real-world testing scenarios, assess component interactions, and validate application behavior, thereby enhancing your Power Apps testing capabilities.

However, what if we want to call an API but add a special header for alternative routing? What if you want to actually call the API and change the resulting output?

Playwright allows different types of routing, such as the one highlighted in *Figure 11.11*, through different potential routes, based on the test plan defined. However, the Test Engine version at the time of writing does not support those different approaches.

We have made a fork – that is, a source code change in **Test Engine** – and included the new version in the book repo to test the aforementioned scenarios.

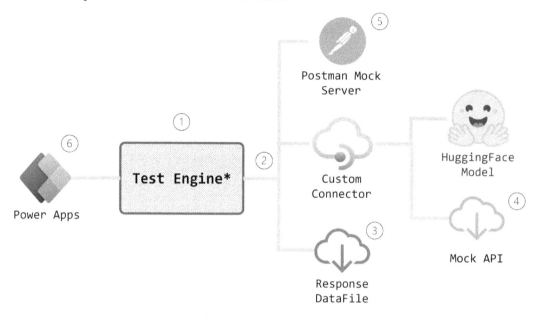

Figure 11.11 – A forked Test Engine flow for different routes

We will use in the third use case (*rerouting the request*) the forked Test Engine version that adds these alternative options to showcase different mocks:

1. First, as defined in the test file, the Test Engine runs the different steps, automating the execution of Power Apps in a selected browser.

2. For the example in *Figure 11.11*, where the app calls the **HuggingFace Model API** to run the AI model, the Test Engine will behave differently, based on some parameters added to the test plan.

3. The default Test Engine behavior will be to intercept the AI call, make a different route, and return the contents of a response **DataFile**. In Playwright, this is called *fulfilling the request*, as previously mentioned.

4. Conversely, we can use an API management service, such as **Azure API Management** (**APIM**), used in many connectors (such as the *weather* component). The parameter will maintain the route to the service but share the request to act as a mock API. In Playwright, this will be run as a *continuation of the request but add new headers*.

5. Instead of using a response file, we want to use a mock server where we can dynamically change the response. I will use Postman – an API platform for building and using APIs that have mock capabilities. How would Test Engine behave in this case? It will route to a different URL, based

on the parameters, and serve the response defined by Postman. In this case, the Playwright will *reroute the request and add new headers.*

6. Finally, the Power Apps app will get the response without noticing the change.

> **Note**
>
> The version of Test Engine used in this chapter is based on the 1.0.5-preview version, published on May 15, 2023. The changes are held in the fork at `https://github.com/carloshm/PowerApps-TestEngine`, while the file used is located in the book repository for easy use and download, as mentioned earlier.

Now that we have presented the scenarios, let's dive into the mock testing technique.

Starting with mock testing

In essence, mock testing is a technique where we simulate the behavior of real dependencies in our applications. Dependencies, in this case, refer to any external systems, services, or modules that our application relies on. These can include databases, network calls, APIs, and connectors.

Mock testing becomes particularly useful when testing Power Apps that make calls to those services. Rather than making actual calls to these external systems, which could have unwanted side effects, mock testing allows an application to be tested without modification and without invoking these dependencies. This gives us a level of control over the behavior of these dependencies and lets us isolate our application code for testing.

Why mock testing?

Mock testing has several benefits, including the following:

-   It improves the speed of testing, since there's no need to wait for responses from external systems
-   It allows us to simulate and test specific scenarios, even edge cases, which might be challenging to reproduce with real dependencies
-   It reduces the impact of unstable dependencies on your testing, allowing you to maintain progress even if a dependent service is down

So, let's see how we can add to our Power Apps test mocks.

Mock testing in Test Engine

Test Engine is a tool in Microsoft Power Platform that allows users to automate and manage testing for their Power Apps. This tool also supports mock testing.

When we talk about mock testing in the Test Engine context, it refers to creating mocks of network calls. This way, it imitates the dependencies that your Power App relies on. It's an excellent way to ensure your Power Apps behave as expected under various conditions, without the need to alter the app itself or deal with potential fallout from external services.

Let's recall some of what we learned from *Chapter 7, Working with Test Engine*. Check it out now to reacquaint yourself with some of the concepts. We will present here the specific elements to add to the **YAML** representation for mock testing, which you previously reviewed as `NetworkRequestMocks` schema at `https://github.com/microsoft/PowerApps-TestEngine/blob/main/docs/Yaml/test.md#networkrequestmocks`.

Network mock files

When you need to include network request mocks, you will consider two files. The first one is `testPlan.fx.yaml`, which is used to describe the tests to run. In the test YAML file, the `mock` description will be inside the `testSuite` heading as a `networkRequestMocks` section, where the configuration will be added for each request, we want to intercept and manage.

The following table presents the properties for each request to include in the file, available in the document at `https://learn.microsoft.com/en-us/power-apps/developer/test-engine/yaml`:

Property	Required	Description
requestURL	Yes	This is the request URL that will get mock response. Glob patterns are accepted
responseDataFile	Yes	This is a text file with the mock response content. All text in this file will be read as the response
Method	No	This is the request's method (GET, POST, etc.)
Headers	No	This is a list of header fields in the request in the format of [fieldName : fieldValue]
requestBodyFile	No	This is a text file with the request body. All text in this file will be read as the request body

Table 11.1 – The NetworkRequestMocks properties section for testplan

The second file is `response.json` (for each example, the filename will match the service name), which is referenced in the `responseDatafile` and will be served as the mocked response.

The main information to include is `requestURL`. As we saw in earlier chapters, a quick trick to understand the request made under the hood is to open **Power Apps Monitor** and view the network or service request, allowing you to define the right glob pattern.

What is a glob pattern?

A *glob pattern* is a way to specify sets of filenames using wildcards. It's commonly used to match file paths for our scenario URIs.

Here are four straightforward examples:

- `http://example.com/*.jpg`: Matches all `.jpg` images on `example.com`
- `http://example.com/images/*/thumbnail.png`: Matches `thumbnail.png` inside any direct subdirectory of images
- `http://example.com/docs/??.pdf`: Matches PDF files in docs with exactly two characters before the `.pdf` extension, such as `01.pdf` or `ab.pdf`
- `http://example.com/data/[a-c]*`: Matches URLs in data that starts with a, b, or c

Connectors in **Power Platform** usually use **Application API management**, located at `https://azure-apim.net/`. This means that your app may have multiple services under the same domain name, and it is critical to differentiate which mock configuration points to the right URL. For that scenario, the glob pattern comes to the rescue.

Using advanced mocking scenarios in the forked Test Engine

We extended the Test Engine behavior in our fork, but *how could you activate the new mock behavior we described earlier?* It is simple – through extended headers. For each strategy, you will add an extended header such as the following:

1. The default Test Engine behavior will be to *fulfill* the request with the `response.json` file. You can add a header with the name `x-mock-type: 0` or just stay without it.

2. If you want to maintain the route to the service but add metadata so that the service behind could act as a mock API, for example, the forked Test Engine will *continue* the request but *add new headers*. This will be done through `x-mock-type: 1` header.

3. Finally, instead of using a response file, we want to use a mock server where we can dynamically change the response. Test Engine will route to a different URL, based on the parameters, and serve the response defined by the mock server.

The request will be rerouted, and new headers will be added. This is possible through the `x-mock-type: 2` header, and the `x-mock-server-url` header value should be set to the mock server URL.

Let's see the app being tested in more detail.

Testing mock scenarios with a Power Parse app

A Power Parse app will be the demo app that will be run in all tests. Let's describe its main screens briefly. You can see the home page describing how to use it in *Figure 11.12*.

Welcome to *Power Parse!* 🎉

This is a sample app for the Automate testing for Power Apps book.
The book presents some of the Power Apps features, Test Engine and Test Studio capabilities and some testing concepts overall.

The app uses Artificial Intelligence 🧠 to summarize 🗒, categorize ◎ and answer some question 🐵 you may have about life in general 🙂 and specifically from content included in the app.

How could you use this app?

1. ⚙ First, configure the AI endpoints you are going to use. The app once deployed present between Azure OpenAI Service Davinci model and Hugging Face Bloom options.
2. ✚ Second, I recommend to add some text or at least look existing data. Knowledge base text and prompts (stored in a SharePoint List and in a dataverse table).
3.   Finally, start asking questions!

Figure 11.12 – The Power Parse home page with the user guide

Once you configure the Power Parse app, and use or create the prompts and content, it will send a request to Azure OpenAI Service or the Hugging Face BLOOM Model, getting a generative AI response based on the prompt and content sent. Basically, it will send a **metaprompt** as a result of the concatenation of both pieces of information – `Prompt or instruction\n\nContext Content`. *Figure 11.13* shows the screen where choosing a prompt and its corresponding context generates an email text with the selected model.

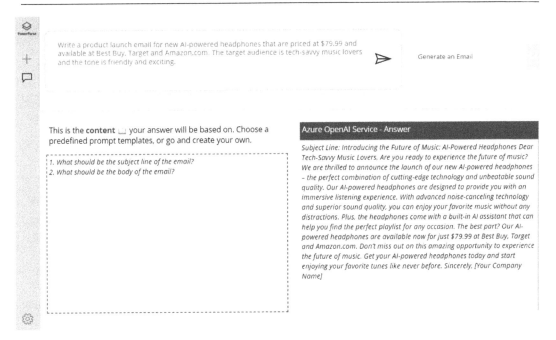

Figure 11.13 – The Power Parse generative AI question screen

We created this app to include all the mocks needed and all mock strategies, including the following:

- A SharePoint List storing context AI content

- Dataverse table storing generative AI prompts, with the instructions

- Information that is sent to API services and consumed through Power Automate Cloud Flows as connectors

In the upcoming sections of this chapter, we will be diving deeper into how to mock specific dependencies such as SharePoint services, Dataverse, and Power Apps connectors. By the end of this journey, you'll be well-equipped to handle mock testing in Test Engine and improve your testing efficacy.

Running the following tests will require following these steps, as described in *Chapter 7*:

1. Import the **Power Parse** solution (as mentioned in this chapter's technical requirements)

2. Update the values of `config.dev.json` with the `environment`, `tenant`, and `testplan` files and the `output` directory.

3. Set the username and password for the test execution.

4. Review the `testPlan.fx.yaml` file for each test and make any required changes. The different tests are located at `https://github.com/PacktPublishing/Automate-Testing-for-Power-Apps/blob/main/chapter-11/` in the `testplans.zip` file.

5. For the SharePoint and Dataverse scenarios, you can use the original Test Engine from *Chapter 7*; conversely, for the third example, the Power Apps connectors, you can run the forked version mentioned earlier – `PowerAppsTestEngine_chapter_11.zip`.

6. Finally, inspect the test results, logs, and recordings.

Now that we know the different routes and the format file we can use in our solution, let's dive deep into each example.

Mocking a SharePoint Service from Microsoft 365

Let's take a simple scenario – imagine you have a Power App that fetches data from a SharePoint service. In a real-world scenario, every time you run a test, the app would make a call to the SharePoint service, fetch data, and consume resources, both in your app and the SharePoint service. It might also be subject to network latencies. With mock testing in Test Engine, you could mimic the SharePoint service response. This way, you would save resources, avoid network latencies, and ensure your app behaves correctly when it receives data from SharePoint. As described previously, *Figure 11.14* shows that, in this case, we will serve the response DataFile for the request through a path (**3**), once the request (**2**) is detected in the SharePoint list.

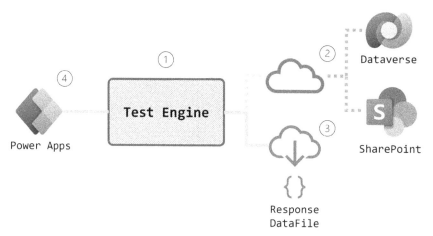

Figure 11.14 – A SharePoint service test flow diagram

After following the previously listed configuration steps, you will have the following:

- A command prompt, configured and opened in a folder with `PowerAppsTestEngine.exe` (the forked version of Test Engine), and a `testplans` folder with the different test plans

- The `config.dev.json` file pointing to the SharePoint test – that is, `"testPlanFile": "./testplans/SharePoint/testPlan.sp.fx.yaml"`

> **Important**
> You should run the forked version of Test Engine on the same computer you configured previous Test Engine examples; conversely, install the Playwright dependency.

In each `testplans` folder, you will have two different files with the raw data from Power Apps Monitor – `PowerParse-Monitor-XX-Content-Interaction.json` file where you can check info such as URL and headers, and `response-XX.json`, where you can check the real data returned from the service. That information will be used to define the mock data.

For example, in the Power Apps Monitor file, when checking for the SharePoint URL you will find this value, `https://europe-002.azure-apim.net/invoke`, which would be the best `requestURL`? We could follow these rules unless we are intentionally testing a specific location (for example, `europe` or `unitedstates` values):

- **Generalizing region information**: value `europe-002` changed to `*`

- **Including information to discriminate different endpoints of the same service**: SharePoint, Dataverse, and Connectors could have the same value of `requestURL` and `method` for all requests, so we will add `x-ms-request-method` and `x-ms-request-url` in headers to identify different requests. Example `/apim/sharepointonline/` that identifies the SharePoint list endpoint.

So, for `requestURL`, we will leave the following glob pattern: `https://*.azure-apim.net/invoke`.

We will then add `x-ms-request-url` with the value of the API SharePoint call as a header. This information will be included in the `testPlan.fx.yaml` file describing the test to fulfill the request; we will highlight only the `networkRequestMocks`:

```
networkRequestMocks:
    - requestURL: https://*.azure-apim.net/invoke
      headers:
        x-ms-request-url: /apim/sharepointonline/
a95e26ffb84e4a71842ac2aceba77d28/datasets/https%253A%252F%252F20mpt1.
sharepoint.com/tables/3e6176d8-1c33-4176-aced-8b4ed165f00b/
items?%24top=500
      responseDataFile: ./SharePoint/response-sp-ai-content.json
```

`response-sp-ai-content.json` is the file describing the content to fulfill the request with the response data from the SharePoint list. You can check from the `Monitor` file that it is the content of the `body` tag. In the next codeblock, you can see a short version of its content:

```
{
    "value": [
        {
```

```
        "ID": 10,
        "Title": "Generate a mock Email",
        "Description": "What should be the subject line of the
email? \n What should be the body of the email?"
    },
    {
        "ID": 11,
        "Title": "Generate a mock Quiz",
        "Description": "A neutron star is the collapsed core of a
massive supergiant star, ... "
    },
    {
        "ID": 12,
        "Title": "Classify mock Text",
        "Description": "Headline 1: Ruth, Charo and Nicolas are
Cooking Up a New Kind of Perfection.
    },
    {
        "ID": 13,
        "Title": "Key points from mock report",
        "Description": "# Start of Report\nRevenue increased $7.5
billion or 16%. End of Report"
    }
]
}
```

We recommend you change data from the response file, so you validate the mock response actually works and shows your changes. For example, the title value of the items.

> **Note**
>
> Remember to check the user configured to run the test has permission to run the app and to authorize app access to any service on your behalf. Otherwise, you will see a user prompt in the recording.

You could get your own response files by downloading them from **Power Apps Monitor**, allowing you to test different data configurations without actually changing the **SharePoint list** data.

You could review the results of the test afterward in the GitHub repo. *Figure 11.15* presents a sample of the execution with (right) and without (left) mock:

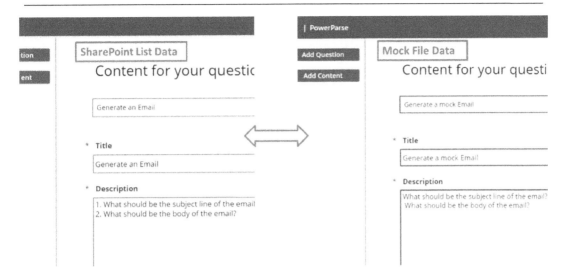

Figure 11.15 – Side-by-side comparison (left) app without mock, (right) with mock

Let's continue with the other strategies!

Mocking Dataverse in your Power Apps

All the information from the Power Parse app could have been read from SharePoint, but for the sake of this example, we have added a different data source to showcase the network request from Dataverse tables. In a real-world scenario, you may want to make changes to the data in a test environment; however, this would lead to potential inconsistencies and the need to set up a data baseline for each test. As long as you maintain the right state, you can mock requests with response files as well. *Figure 11.16* shows that, in this case, as described previously, we will serve the response DataFile for the request through a path (**3**), once the request (**2**) is detected in the prompts table.

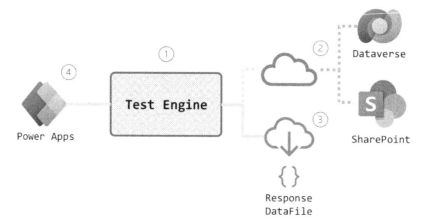

Figure 11.16 – The Power Parse generative AI question screen

Repeat the same previous steps as for the SharePoint test. Then, for `requestURL`, we will leave the following glob pattern: `https://*.crm4.dynamics.com/api/data/v9.0/$batch`. In this case, we may add a `requestBody` file with information, but for our scenario, there is no need to add additional headers.

This information will be included in the `testPlan.dv.fx.yaml` file containing the test to fulfill the request; pay attention to the glob pattern in the next code detail.

In `testPlan.dv.fx.yaml` file describing the test to fulfill the request, we will highlight only the `networkRequestMocks`:

```
networkRequestMocks:
  - requestURL: https://*.crm4.dynamics.com/api/data/v9.0/$batch
    headers:
      x-mock-type: 0
    responseDataFile: ./Dataverse/response-dv-ai-prompt.json
```

`response-dv-ai-prompt.json` is the file describing the content to fulfill the request with the response data from the Dataverse store. You can check in the `Monitor` file that it is the content of the body tag. In the next code you can see a short version of its content:

```
{
    "value": [
        {
            "cr87f_promptsid": "a398bc58-...",
            "cr87f_description": "Generate the mock Email",
            "cr87f_name": "Write a product launch email for new AI-
powered headphones..."
        },
```

```
    {
            "cr87f_promptsid": "999c68a7-0b19-...",
            "cr87f_description": "Generate a mock Quiz",
            "cr87f_name": "Generate a multiple choice quiz from the
text below. ..."
    },
    {
            "cr87f_promptsid": "7fab9ebe-...",
            "cr87f_description": "Classify mock Text",
            "cr87f_name": "Classify the following news headline into 1
of the following categories..."
    },
    {
            "cr87f_promptsid": "5620b4e0-...",
            "cr87f_description": "Key points from the mock report",
            "cr87f_name": "Below is an extract from the annual
financial report of a company founded by Marta and Raul. ..."
    }
    ]
}
```

You could review the results of the test afterward, in the GitHub repo. *Figure 11.17* presents an example of the execution with (right) and without (left) mock:

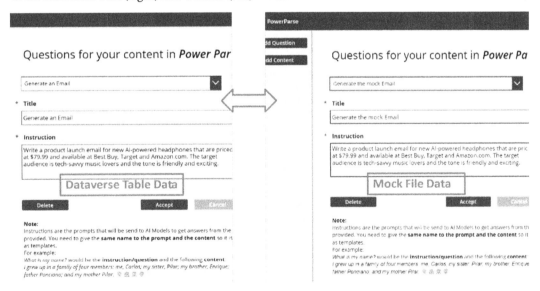

Figure 11.17 – Side-by-side comparison (left) app without mock, (right) with mock

Let's close with the final strategy, rerouting.

Mocking a Power Apps connector

Let's take a simple scenario – imagine you have a Power App that fetches data from an AI model such as HuggingFace or Azure OpenAI. In a real-world scenario, every time you run a test, the app would make a call to those APIs, fetch data, and consume resources, both in your app and the AI Model Service. It might also be subject to network latency. With mock testing in **Test Engine**, you can mimic those service responses. This way, you would save resources, avoid network latency, and in the AI scenario, ensure your app receives predictable responses. *Figure 11.18* shows that, in this case, as described previously, we will serve the Postman mock response, instead of a DataFile or the real service for the request through the path (**3**), once the request (**2**) is detected.

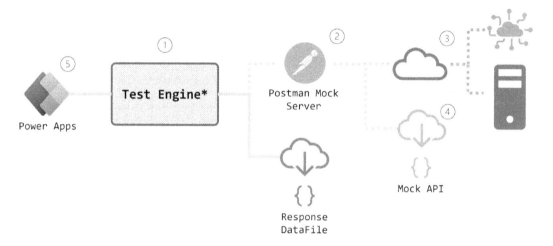

Figure 11.18 – HuggingFace AI Model flow diagram

Once you have a free Postman account, you will create two different objects:

- Create a new collection (**1**), and give it the name `Power Apps`. Then, add a new request (**2**) and add the following data to the request:

  - Name: Bloom (**3**)

  - URL info: POST `Method`, and HuggingFace BLOOM Model URL and `Authorization Bearer` token (gathered in the technical requirements section) (**4**)

- And a content body with a sample content (**5**), as shown in *Figure 11.19*.

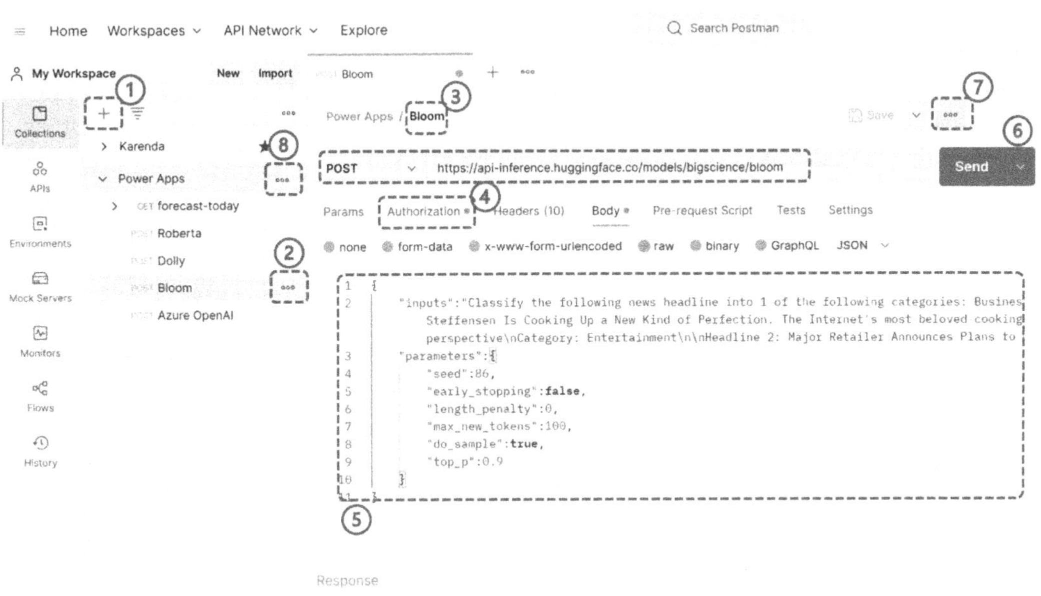

Figure 11.19 – Collection configuration in the Postman app

- You can check the request by clicking **Send** (**6**), and save an example of the response as a mock (**7**) from the real service request. You can see how to create examples at https://learning.postman.com/docs/sending-requests/examples/.

- Save the collection.

The following code is used as the body text of the request:

```
{
    "inputs":"Classify the following news headline into 1 of
the following categories: Business, Tech, Politics, Sport,
Entertainment\n\nHeadline 1: Donna Steffensen Is Cooking Up a New
Kind of Perfection. The Internet's most beloved cooking guru has a
buzzy new book and a fresh new perspective\nCategory: Entertainment\n\
nHeadline 2: Major Retailer Announces Plans to Close Over 100 Stores\
nCategory:",
    "parameters":{
        "seed":86,
        "early_stopping":false,
        "length_penalty":0,
        "max_new_tokens":100,
        "do_sample":true,
```

```
        "top_p":0.9
    }
}
```

- Create a mock server through a mock collection (**8**); this will allow you to send requests to a defined mock server URL provided by Postman. Give the mock server a name, leave everything else as default, and click **Create Mock Server**. This will give you a URL to configure in Test Engine.

- You could test in Postman, creating a new request with the same method and URL endpoint (changing the domain), as shown in *Figure 11.20*

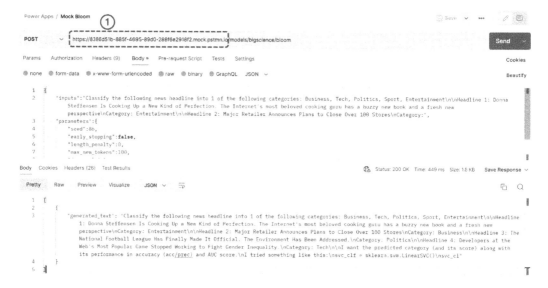

Figure 11.20 – A request sent to the mock server in the Postman app

We are now prepared to use the new URL in the test plan, configure a new header to reroute the request to Postman, and get the results.

The `testPlan.hf.fx.yaml` file contains the test to reroute the request. We will highlight only the `networkRequestMocks`:

```
networkRequestMocks:
  - requestURL: https://*.azure-apim.net/invoke
    headers:
        x-mock-server-url: https://8386d51b-885f-4695-89d0-
288f6e2918f2.mock.pstmn.io/models/bigscience/bloom
        x-mock-type: 2
        x-ms-request-url: /apim/logicflows/08a86e4d804ad550bc97e957c
028a1fb-b980fa5e8e32f9ad/triggers/manual/run?api-version=2015-02-01-
preview
```

```
        responseDataFile:  ./HuggingFaceBLOOM/response-huggingface-ai.
  json
```

In this case, the custom header `x-mock-type:  2` instructs the forked Test Engine executable to reroute the request with the response data from the Power Apps connector identified by `x-ms-request-url:/apim/logicflows/08a86e4d804ad550bc97e957c028a1fb-b980fa5e8e32f9ad/triggers/manual/run?api-version=2015-02-01-preview` to the postman mock server identified by `x-mock-server-url:https://8386d51b-885f-4695-89d0-288f6e2918f2.mock.pstmn.io/models/bigscience/bloom`, getting the response from *Figure 11.20*.

You can review the results of the test afterward, in the GitHub repo. *Figure 11.21* shows how you could can the actual request in your Postman Mock Server.

Figure 11.21 – Mock request in the postman Mock Server

That's it! We have come to the end of the chapter reviewing the different mock strategies when managing network requests. Remember to check the GitHub repository for additional examples and details.

Summary

In this chapter, we explored the crucial concept of mock testing, with a special emphasis on its application in Test Engine. We covered essential topics such as understanding and mock testing and how to apply it, and the emulation of dependencies such as SharePoint Service from Microsoft 365, Dataverse, and Power Apps connectors. You've now become proficient in mock testing in Test Engine, gained the ability to perform tests when your Power Apps have dependencies on external services, and minimized potential disruptions from these services. With these skills, you're better equipped to enhance your Power Apps testing processes.

The following chapter will delve deeper into the telemetry of Power Apps, a vital aspect of testing. This involves collecting and analyzing usage and behavior data, which provides valuable insights to debug and understand user interaction, leading to improved testing strategies.

12

Telemetry and Power Apps

In the previous chapter, we ventured into the crucial concept of mock testing, focusing primarily on its implementation in Test Engine. We examined essential facets such as understanding mock testing and emulating dependencies, which included the SharePoint service from Microsoft 365, Dataverse, and an example of AI services in Power Apps.

In this chapter, we shift our focus to a crucial aspect of testing—the collection and analysis of telemetry data from Power Apps. This allows insightful debugging and a better understanding of user behavior, hence contributing to effective testing. We will navigate through the following areas:

- **Power Apps analytics and Monitor**: Here, you will gain insights on how to retrieve and interpret data from Power Apps analytics and Monitor

- **Dataverse analytics**: You will learn how to gather information from Dataverse, one of the most widely used services in Power Apps

- **Application Insights in Power Apps**: This section equips you with the skills to extract and interpret data from Application Insights utilized by Power Apps

- **Getting insights on your data**: You will discover how to query telemetry traces and logs in Application Insights with **Kusto Query Language** (**KQL**)

By the end of this chapter, you will have gained comprehensive knowledge about telemetry in Power Apps, understood how to interpret this data, and identified testing scenarios using this information. This knowledge will prove invaluable in debugging and improving your Power Apps.

Technical requirements

To follow the examples in this book, you must publish any application shared before in the book, or one of your own, and use it so that it has activity. For the **Application Insights** section, you need an Azure subscription to configure the service.

Access to a functioning **Power Platform environment** is necessary in order to install the provided examples or create new ones. The Power Platform developer environment appears to be the most suitable solution for this purpose. In previous chapters, you will see steps about how to obtain access to such an environment.

We will utilize two platforms:

- **Power Platform admin center**: If you set up a developer environment, you will have an account with administrator permissions. On the other hand, Power Apps analytics at the app level was described in the previous chapter. Here, we will focus on the admin center. Two of the analytics services need this access level:

  - **Power Apps analytics**

  - **Dataverse analytics**

- **Application Insights** in **Azure Monitor**: This is broadly used for monitoring and diagnostics. It stores app telemetry data in Azure Monitor logs. You can follow the steps from `https://learn.microsoft.com/en-us/power-apps/maker/canvas-apps/application-insights`.

In a nutshell, the main steps to use Application Insights are as follows:

1. Create or use an Azure account, as described in the docs at `https://learn.microsoft.com/en-us/dotnet/azure/create-azure-account`.

2. On the home page, click on **Create a resource**, search for `application insights`, as shown in *Figure 12.1*, and click on **Create**:

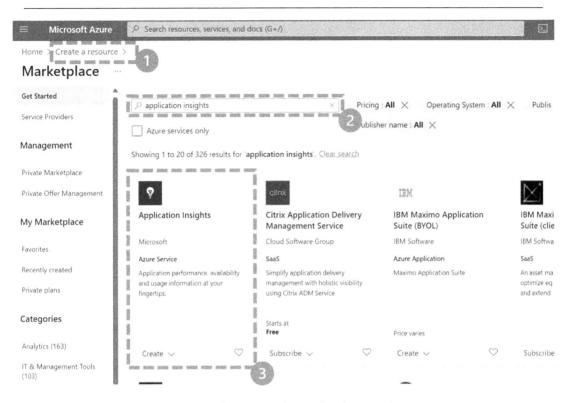

Figure 12.1 – Application Insights marketplace creation page

3. Once you click on **Create**, the process will require you to add some more information, including creating a new resource group, choosing an Azure region near you, and more. For the rest of the options, leave them at their defaults and use an instance name for the service. *Figure 12.2* presents the information to be completed:

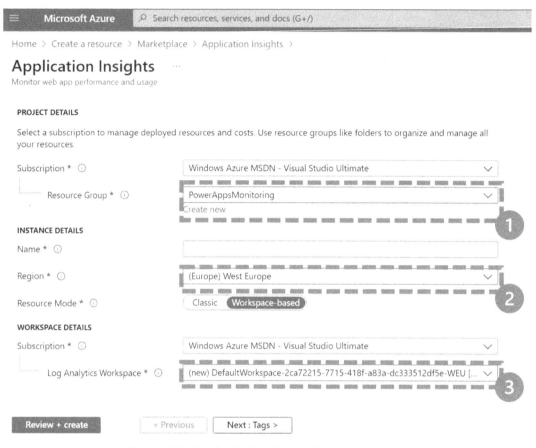

Figure 12.2 – Application Insights detailed creation page

4. Finally, once your Application Insights instance has been created, you will need the **Instrumentation Key** value to be included in your apps so that the app telemetry is stored in the service. You will need to navigate to **Overview** in the left menu of the created Application Insights service and get the **Instrumentation Key** value as it appears in *Figure 12.3*:

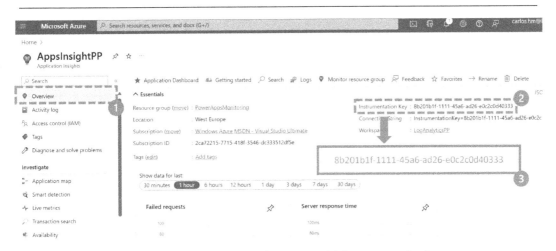

Figure 12.3 – Application Insights service with instrumentation key

Azure Application Insights is based on Azure Data Explorer technology. To test KQL, we will explore some of its capabilities in a free environment, if you don't have access to Azure. You can refer to the *How do I create a free personal cluster?* section of `https://detective.kusto.io/faq`.

If you don't want to use the out-of-the-box services, once you have access to the two environments, Power Platform admin center and Azure Application Insights, you can move on to the next sections.

Power Apps analytics

Telemetry offers crucial insights into the internal workings of an application and the services used. In the absence of such data, understanding what is going on in your app would essentially be like going blind. Telemetry empowers you to identify and monitor specific operations, thereby helping you understand if the system is operating as expected or if certain aspects are detrimentally impacting it.

When you augment applications utilizing custom components or introducing server-side logic, telemetry becomes your guide. It allows you to perceive the influence these elements may impose on performance and presents opportunities for optimization, inclusive of design modifications if necessary.

Moreover, telemetry plays a pivotal role in monitoring overall performance trends, enabling you to take preemptive actions to manage them instead of simply responding to user incidents.

As we mentioned in *Chapter 4, Planning a Testing Phase in a Power Apps Project*, there is information gathered at the app level and presented through some reports, but this is available at the environment level for admin users. This gives you a series of dashboards and reports on environment-level usage, errors, and performance, a place where you can more easily browse and change between environments to look into analytics data.

You can see the different elements in *Figure 12.4*. Once you navigate to the Power Platform admin center and click on the **Analytics** section on the left menu, you will see the **Power Apps** panel (*1*). The dashboard will allow you to change between environments (*2*), such as testing and production environments, and for each environment, you will get different areas to check. The maximum retention is 28 days, as is highlighted in the screenshot, so the other alternative will help you move beyond that limit.

In the same screenshot, you can see the different options for the top menu. Let's list them all:

- The **Overview** option will show you tenant-level reports, as long as they are enabled by the tenant admin

- For the **Environment View** option, you will have five different areas: **Usage**, **Location**, **Toast Errors**, **Service Performance**, and **Connectors**

- Each chart will allow you to do the following:

  - Export its data

  - Show it as a table in detail

  - Filter with side properties values such as **Device**, **Version**, or **Country**, and define the number of days the report shows (28, 14, or 7 days)

  - Maximize each chart for a full screen view:

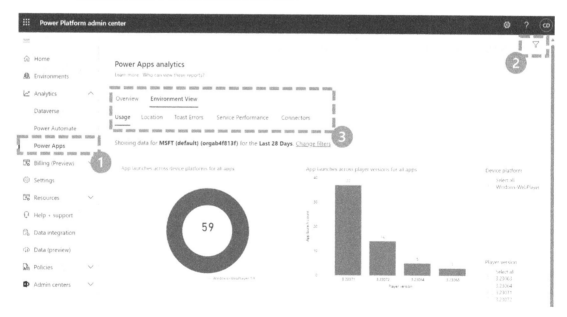

Figure 12.4 – Power Apps analytics: Power Platform admin center view

As an example, for our Power Parse app that creates content with generative **pre-trained transformer (GPT)** models, specifically Azure OpenAI models, a change was made to the API key to query the service. The API key is part of an environment variable, so you need to publish the change or the application will not work. In this hypothetical scenario, the rise in the number of errors in the performance report alerted us about an issue with the `logicflows` calls highlighted in the service performance view shown in *Figure 12.5*, where you can see 500 errors and unsuccessful requests:

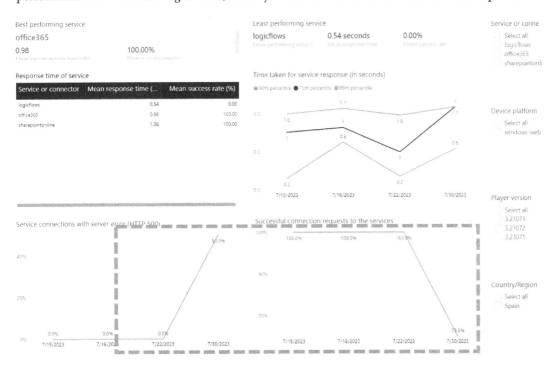

Figure 12.5 – Power Apps analytics: service performance view errors

Tracking these reports should be your minimum baseline when managing your solution, as it will give you initial information about your app moving from reactive to proactive, in the experience of your users. Adding this into Application Insights and alerting and running tests to validate the service in all environments will empower you with the best insights. We will check this same case later.

Monitor

The **Monitor** dashboard provides a platform for you to inspect properties for every app event. It offers insights into the actions triggered when a control is selected, the process's timespan, and the result of the operation.

Figure 12.6 provides a representation of the **Monitor** dashboard's appearance. After the screenshot, you will find a breakdown detailing the information involved in each column of the dashboard. Once you select a specific row, a **Monitor Properties** panel appears. Each event will feature a **Details** and **Formula** tab, and when an event initiates an HTTP request/response, additional tabs for **Request** and **Response** will be visible:

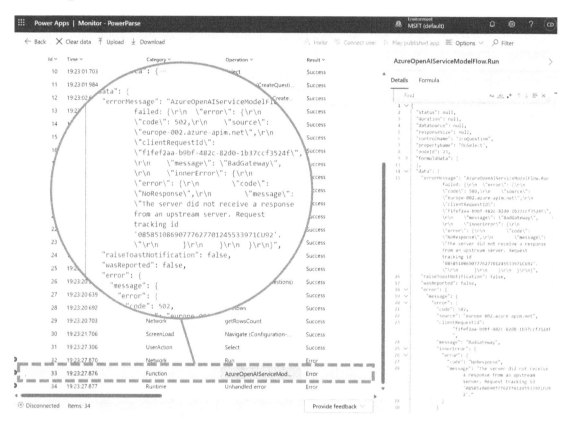

Figure 12.6 – Power Apps Monitor: logic flow view

There are two ways to activate Monitor:

- When you edit a canvas app, go through the **Advanced Tools** menu and open **Monitor**.

- The other option is the other way around. First, open **Monitor** from an app or solution view in Power Apps, and then click **Play** for the associated published app. Remember from previous chapters that you can connect to someone else's session in a collaborative way.

In *Figure 12.6*, you can see an error related to the Azure OpenAI service (ID *33*), which we will talk about later in the chapter.

As we will query Monitor logs later, we will describe the type of information available and how it is mapped to the download file exported through the UI.

Requests associated with user interaction with the app will be logged with the following information (we will take the number 32 as an example):

- **Id**: [32], an incremental number reset for each session.

- **Time**: [19:23:27.870], the timestamp of the start of the request.

- **Category**: [Network], a literal text describing the category of the request. Other examples are ScreenLoad, UserAction, Function, Runtime, and Performance.

- **Operation**: [Run], a literal text with the operation executed. Other examples are Select, getRows, Navigate, getRowsCosunt, LoadScreen, and Unhandled error.

- **Result**: [Success/Error], a literal text with the type of result—a success or an error.

- **Result info**: [Bad Gateway], a description of the result—in this case, a propagated error from the Azure OpenAI service. For getRowsCount, it will describe the result action: Requested 500 rows. Received 5 rows.

- **Status**: [502] Status HTTP code.

- **Duration (ms)**: Duration of the request in milliseconds.

- **Data source**: [AzureOpenAIServiceModelFlow], a literal with the name of the data source—in this case, a Power Automate flow name.

- **Control**: [icoQuestion], the name of the control that invoked the action.

- **Property**: [OnSelect], the property of the control.

- **Response size**: The size in bytes of the response.

All requests will have a detailed view and formula, and if it is a network request, the corresponding request/response.

When you download the log, by clicking on the **Download** option of the top menu, a file named PowerAppsTraceEvents.json will be downloaded. It will have the following format:

```
{
  "Version": 2,
  "SessionId": "a96f6ec0-2efd-11ee-90c8-cd5a8123a018",
  "Messages": [],
  "Config":[]
}
```

The content of each message will be found in the following schema:

```
{
    "time": 1690737807870,
    "category": "Network",
    "name": "Run",
    "logLevel": 2,
    "data": {
      "context": {
        "entityName": "icoQuestion",
        "propertyName": "OnSelect",
        "nodeId": 23,
        "id": 655,
        "diagnosticContext": {
          "dataOperation": {
            "protocol": "rest",
            "operation": "Run",
            "dataSource": "AzureOpenAIServiceModelFlow"
          }
        }
      },
    },
  "request":{},
  "response":{}
}
```

With this information, you will be able to drill down into a full view of what is going on in your app, including user actions, network requests, Dataverse activity and performance, and more. You can use any other analytics tools, such as your **Kusto cluster**, **Microsoft Excel**, and **Power BI**, for this.

It is time to review Dataverse analytics as part of the monitoring reports available on the platform.

Dataverse analytics

An app-level report is not available for Dataverse. It is more commonly used in model-driven apps than in canvas apps, as it is used as the default data source, but it is available for any app. Dataverse is a secure and scalable data platform that's integrated into Power Platform; it allows users to store and manage data used by business applications. It is also available at the environment level for admin rights-level users. This gives you a series of dashboards and reports at the environment level for usage, plugins, and API calls—a place where you can browse and change date ranges and environments to look into analytics data.

You can see the different elements in *Figure 12.7*. Once you navigate to the Power Platform admin center and click on the **Analytics** section on the left menu, you will see the **Dataverse** panel (*1*). The dashboard will allow you to change between environments (*2*)—testing and production environments—

and for each environment, you will get different areas to check (*3*), such as active users or API calls, with the same retention limit as before. In this case, you cannot interact with the charts to export data. You can also download defined report data from the top (*4*) to get insights not available in the dashboards (`https://learn.microsoft.com/en-us/power-platform/admin/analytics-common-data-service#download-reports`):

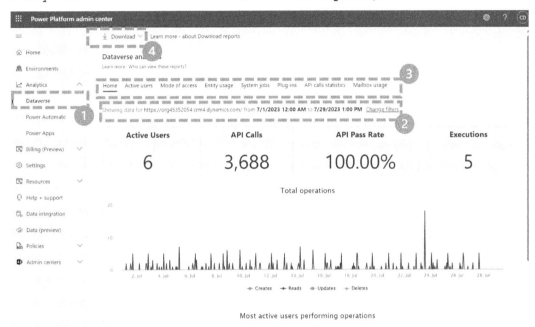

Figure 12.7 – Dataverse analytics: Power Platform admin center view

These reports will give you information focused on the functional view and statistics of the service: app entity usage, API call statistics, or active users. From a canvas app perspective where you will create a Dataverse table to manage some specific data, Application Insights will give you more granularity and control, and its data will be more actionable. Remember that business trace messages from the app could be visible and analyzed in Application Insights.

A good workflow would be this: check **Analytics** in the admin center with out-of-the-box information and monitor the performance of your app through Application Insights; then, if you discover an issue or potential unexpected activities in any environment, use Power Apps Monitor to gather specific information of the issue, import this, and analyze it in more depth with KQL.

Application Insights in Power Apps

Once you have created an Application Insights resource and got the **Instrumentation Key** value, you need to assign it to your Power Apps app. This will be done by selecting the app level in the **Properties** panel, as shown in *Figure 12.8*:

Figure 12.8 – Application Insights instrumentation key in Power Apps app

As soon as you publish the app and use it, all telemetry will go to the Application Insights service.

> **Important note**
>
> When you import or export a solution or Power Apps application file, review the **Instrumentation key** property. You could start sending telemetry information to other tenants or unexpected places. Verify which instrumentation key is being used, and if the process is in place in your **application lifecycle management** (**ALM**) strategy, consider this to be changed through the different environments.

When you navigate to the Application Insights service page, you will find several sections in the left menu to explore its functionality. These features allow you to get a better understanding of the app to diagnose issues and analyze potential feedback from testing. The data is stored in Azure Monitor logs by Application Insights and visualized in the **Performance** and **Failures** panels under **Investigate** on the left pane.

For our apps, the telemetry will be populated in the following tables shown in *Figure 12.10*:

- The `pageViews` table will store page loads

- The `Dependencies` table will store outbound network requests

- The `Requests` table will store Dataverse API incoming calls

I will point out the most useful panels in Application Insights to use for our testing purpose, visible in *Figure 12.10*:

- First, from **Overview**, you could click on **Application Dashboard** and create a dashboard where you could include different visualizations and data.

- Under the **Investigate** section, the **Performance** panel displays the number and average time of each process associated with the application. This data can be utilized to pinpoint which operations impact users the most.

- Also, in the **Investigate** section, the **Failures** panel presents a count of unsuccessful requests and identifies the number of users impacted by each operation for the application. It offers a view into error specifics associated with operations and dependencies, involving both server and browser.

- To check failed transactions such as exceptions or find individual telemetry items, you could use **Transaction Search**.

- Finally, in the **Monitoring** section (the one we are using in this chapter) is the **Logs** panel. Here, you will be able to run queries and browse the different tables, adding queries and their results as pins in dashboards, exporting to other systems such as Power BI or Excel, or you can establish conditions that would trigger an alert whenever a metric surpasses a predefined threshold.

Using Application Insights, we can get more insights from the error we reviewed in *Figure 12.5* about the changed API key error. Looking into the **Failures** panel, we see a spike in *Figure 12.9*. We can go directly for further analysis through the following steps highlighted in *Figure 12.9*:

1. **View in Logs** will jump to the **Logs** panel with all information from the failure's filters and view.

2. The **Logs** panel will present the specific KQL query we can tune to our needs that replicates the information shown previously.

3. We can create a new alert rule with that type of error for future use:

Figure 12.9 – Application Insights API key failure scenario

As you can gather information related to Power Apps analytics, you can also get information from Dataverse telemetry within Application Insights and model-driven Power Apps. Although we are focusing on canvas apps, you can drill down on other applications at `https://learn.microsoft.com/en-us/power-platform/admin/telemetry-events-dataverse`.

Let's go and look inside to find out how to query our telemetry logs.

Querying data in your analytics

As you have seen, there are out-of-the-box experiences such as Power Apps analytics or Dataverse analytics that let you get basic metrics. You can export some data to get specific insights or cross that data from the information store through your app or Monitor logs. But to get full capabilities, the way to go is Application Insights as this is a more complete service for telemetry and specific trace analysis, to diagnose issues or questions about your app.

Data is stored in **Azure Monitor**, where you can analyze it with Log Analytics. It is powered by capabilities from **Data Explorer**, and it uses a specific query language called KQL. To maximize the analysis of the information, it is important to learn about KQL as it is the way you will query the system. Let's get an initial view of it.

Introduction to KQL for Application Insights

KQL is a powerful language for querying, processing, and visualizing data in Azure Monitor Logs and Application Insights. In the context of Power Apps, KQL allows you to delve deep into your app's performance and usage metrics, making it an invaluable resource for optimizing and troubleshooting your Power Apps applications.

The basic structure of a Kusto query involves a data source (`from`), followed by a series of transformations (such as `where`, `summarize`, and `project`), each separated by a pipe (`|`).

To run the examples with data, you can just browse to the **Application Insights** service, then **Monitoring | Logs**, and write and run the query in the editor. The available tables are presented on the left side, so you can check the names and columns, as shown in *Figure 12.10*:

Figure 12.10 – Application Insights editor

Now that we know the elements of the log editor, let's review basic commands for running queries.

Basic commands

The `table` command is the starting point for most queries, as it specifies the data source. In Application Insights for Power Apps, commonly used tables include `traces`, `requests`, `exceptions`, and `customEvents`.

For instance, to query data from the `traces` table, write the following command:

```
traces
```

This command returns all records in the `traces` table.

The `filter` command (also known as `where`) is used to narrow down results based on specified conditions. For example, to find all traces where the severity level is greater than 2, use the following command:

```
traces
| where severityLevel > 2
```

Here, `severityLevel > 2` is the condition used to filter traces.

The `project` command is used to select specific columns in the result. For example, if we only want to see the `timestamp` and `message` columns in our trace data, try this command:

```
traces
| where severityLevel > 2
| project timestamp, message
```

This will return only *timestamps* and *messages* for traces where the severity level is greater than 2.

The `summarize` command is used to aggregate data. For instance, we can count the number of traces with a severity level greater than 2 per day using the following code:

```
traces
| where severityLevel > 2
| summarize count() by bin(timestamp, 1d)
```

The `bin` function here is used to group timestamps by day, and `count()` counts the number of traces for each day.

Visualizations

Once you write a KQL query, the visualization you choose should align with the type of data you're dealing with and the insights you're seeking. Different charts serve different purposes and can greatly enhance your ability to understand and interpret your data.

The `render` operator in KQL is a versatile command used to define the type of visualization to use for displaying the results of your query. It comes in very handy when you're working within a web UI such as Application Insights logs, as it allows you to create charts and graphs directly from your KQL queries.

It's important to remember that the `render` operator is optional. If you do not include a `render` command in your KQL query, your results will be displayed in a tabular format. Additionally, you can always change the visualization after running your query by selecting a different option from the **Chart type** drop-down menu.

The `render` command is typically placed at the end of your KQL query and is followed by the type of visualization you want to use. Next are some examples.

A `timechart` is useful for displaying how a quantity changes over time. For instance, to display the count of requests over the past 7 days, you might use the following command:

```
requests
| where timestamp >= ago(7d)
| summarize count() by bin(timestamp, 1d)
| render timechart
```

A bar chart is great for comparing quantities across different categories. For example, to compare the count of requests by operation name, use the following command:

```
requests
| summarize count() by operation_Name
| render barchart
```

A line chart is perfect for showing trends over time. To visualize the trend of requests over time, use the following command:

```
requests
| summarize count() by bin(timestamp, 1h)
| render linechart
```

A pie chart is useful for showing the distribution of categories within your data. If you want to show the distribution of page views by duration, we'll first categorize requests into different buckets based on duration:

```
pageViews
| where timestamp > ago(120d)
| summarize count() by durationBucket = case(
    duration <= 1000, '0-1s',
    duration <= 2000, '1-2s',
    duration <= 3000, '2-3s',
    '3s+')
```

In this query, we're categorizing our page view durations into four buckets: `'0-1s'`, `'1-2s'`, `'2-3s'`, and `'3s+'`. We then count the number of page views falling into each of these buckets.

Now, as shown in *Figure 12.11*, to visualize this as a pie chart, follow these steps:

1. Access the editor from the left menu in **Monitoring** | **Logs**.

2. After running the query in Application Insights, you will see the query results below the KQL editor.

3. Next to **Results**, after clicking on the **Chart** option, there's an option to select the type of visualization. Click on this and select **Pie** from the drop-down menu.

4. You should see your data displayed as a pie chart, with the size of each section representing the number of requests in each duration bucket:

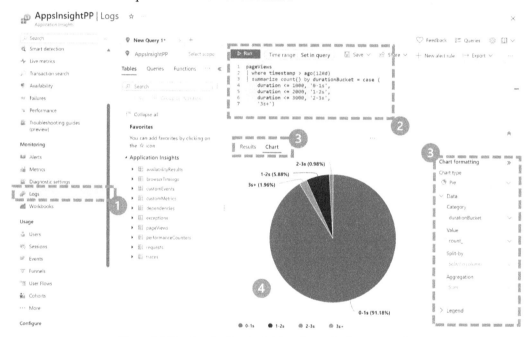

Figure 12.11 – Application Insights editor: pie chart example

If you add | `render piechart` in your editor at the end of the code on a new line, the visualization will directly be a pie chart:

```
pageViews
| where timestamp > ago(120d)
| summarize count() by durationBucket = case(
    duration <= 1000, '0-1s',
    duration <= 2000, '1-2s',
    duration <= 3000, '2-3s',
```

```
    '3s+')
| render piechart
```

Remember—visualization greatly depends on the type of data and the insights you want to derive. A pie chart is suitable when you want to visualize the distribution of categories in your data, but other types of visualizations might be more appropriate depending on your specific needs.

These are just a few of the fundamental KQL commands you can use to query and analyze data in Application Insights for Power Apps. With these basic commands, you can start exploring your data and gaining valuable insights into your app's performance and usage patterns. As you become more comfortable with these, you can move on to more advanced commands and techniques, available at `https://learn.microsoft.com/en-us/azure/data-explorer/kusto/query/`.

> **Note**
>
> Although not related specifically to Power Apps, there is an online game to learn Kusto in a gamified way. Check out `https://detective.kusto.io/`, where you will be part of a detective agency to resolve some of the cases presented. It will validate your KQL knowledge and enable you to use for free Azure Data Explorer, the technology behind Application Insights.

Finally, you could reuse the query and visualization through the export data to Power BI or pin it into the Azure dashboard, as shown in *Figure 12.12*:

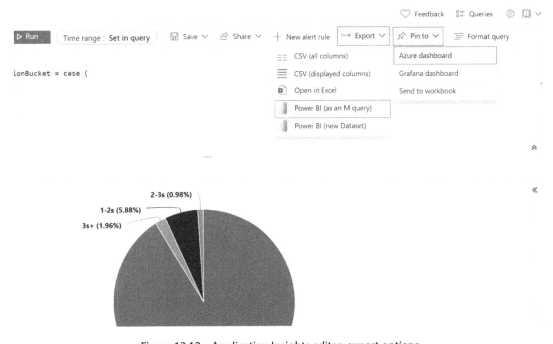

Figure 12.12 – Application Insights editor: export options

Over time, Application Insights is being integrated with more Power Apps scenarios, as well as with the whole Power Platform service and solutions built with it, such as Dynamics 365 apps. So, getting familiar with the service and how to use it makes sense now and in the long term.

Finally, we will briefly showcase the flexibility of KQL for Monitor data logs.

Testing data

As a powerful analytics language, KQL and the original service powering its capabilities, Azure Data Explorer, can be used in some advanced scenarios. It can even be used to consolidate all your data sources such as Application Insights itself, as described at `https://learn.microsoft.com/en-us/azure/data-explorer/query-monitor-data`.

We will describe two scenarios related to information around testing data. In the first example, you were running test cases with Test Studio or Test Engine, and you added a trace message to differentiate from normal operations. You would like to query all traces associated with those tests, including events run in the same sessions, and remove them from the tests. To do this, write the following code in the editor, after adding custom traces from your Power Apps—in our example, `"Power Apps - Testing"`, as you can see here:

```
Traces
| where message == "Power Apps - Testing" and timestamp > ago(30d)
| Project session_Id
| join kind=inner (
    Traces
) on session_Id
```

In the preceding example, we queried the `traces` table in *Figure 12.13*, filtering the entries where we included the Power Fx `Trace` operation with the `"Power Apps - Testing"` text in your Power Apps, and including last month's data filter. We added a `session_Id` column, joining the same table to get additional entries generated by Power Apps in our test case. It appears that for each request, two events were raised: App Start and internal template loader, as presented in *Figure 12.13*:

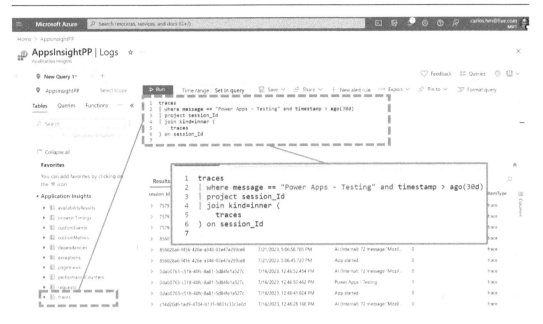

Figure 12.13 – Application Insights: filtering test-case traces

In the second example, in our free Azure Data Explorer cluster, we will import Power Apps Monitor logs. Running Power Apps Monitor and playing the published Power Parse app, you can download the Monitor logs. Then, in your free provisioned cluster, you can follow the three steps presented in *Figure 12.14* to ingest data:

1. Create a new table to store the data.

2. Upload the monitor log files (for example, `PowerAppsTraceEvents.json`).

3.   Validate and import by clicking to ingest data:

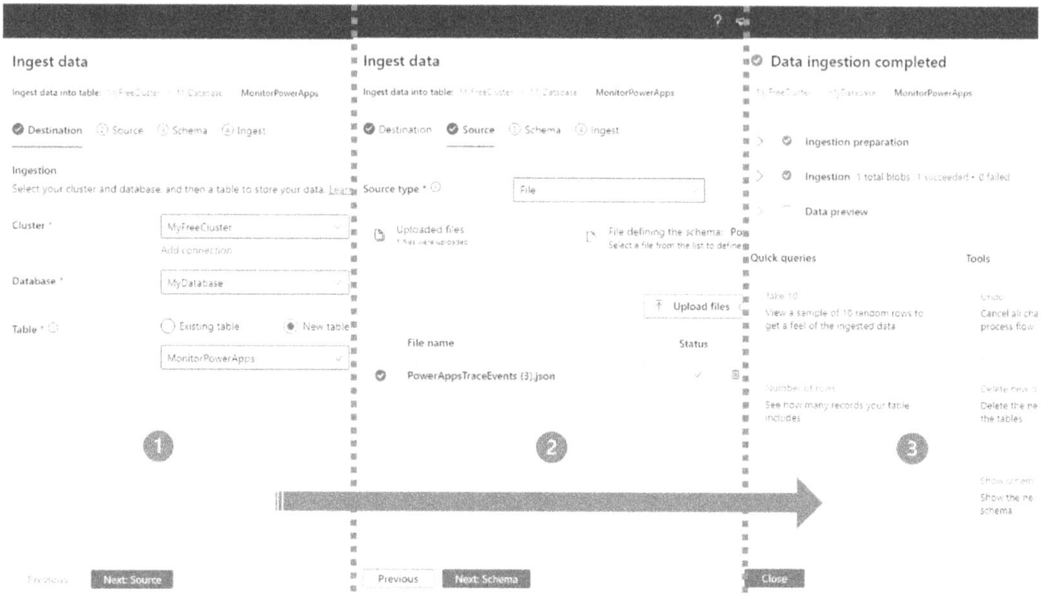

Figure 12.14 – Azure Data Explorer web UI: importing Monitor logs

Once you import the file, you can run a KQL query and explore the powerful possibilities. For this scenario, we use KQL capabilities to query the trace logs and review the time taken for each action on the app.

As we imported the file into our `MonitorPowerApps` table, a single row was inserted, as shown in *Figure 12.15* (there are alternative ways to directly insert items inside a `Messages` array, but we will follow this path for simplicity and to facilitate learning):

⊞ MonitorPowerApps ∨

Version ≡	SessionId	≡	Messages
>	2	a96f6ec0-2efd-11ee-90c8-cd5a8123a018	[{"time":1690737771383,"category":"Network","name":"Launc

Figure 12.15 – Azure Data Explorer web UI: Monitor traces imported

We will use plain KQL to generate the data needed, as presented in *Figure 12.16* and described line by line next:

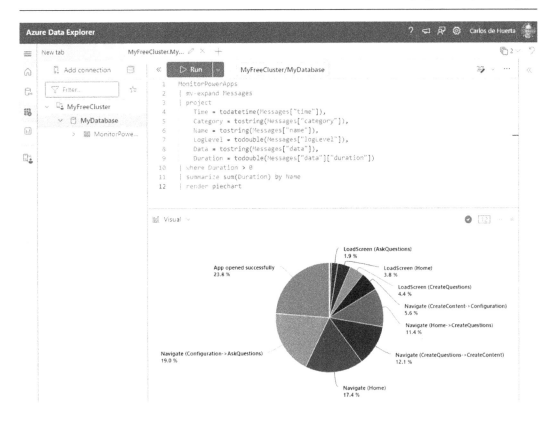

Figure 12.16 – Azure Data Explorer web UI: Monitor trace data processed in a pie chart

1. `MonitorPowerApps`: We define the table to be queried.

2. `| mv-expand Messages`: We expand the `Messages` array column into rows.

3. `| project`: We will extract each of the values in each row into new columns:

   I.   `Time = todatetime(Messages["time"]),`

   II. `Category = tostring(Messages["category"]),`

   III. `Name = tostring(Messages["name"]),`

   IV. `LogLevel = todouble(Messages["logLevel"]),`

   V.   `Data = tostring(Messages["data"]),`

   VI. `Duration = todouble(Messages["data"]["duration"])`

4. `| where Duration > 0`: We filter those rows without duration.

5. `| summarize sum(Duration) by Name`: We group `Operations` by duration.

6. `| render piechart`: Finally, we render a pie chart.

As you start getting familiar with KQL, you will find it very easy to query, in near real time, different perspectives of your test data to look for performance, behavior, or issues.

Summary

This chapter highlighted the pivotal role of telemetry in Power Apps, enabling us to gather and dissect data for improved debugging and understanding of user interactions. We dived into Power Apps analytics, explored Dataverse analytics, harnessed Application Insights in Power Apps, and used these insights to identify testing scenarios. You're now equipped with a thorough understanding of Power Apps telemetry and its application in debugging and enhancing your apps. As we approach the final sections of our enlightening journey, get ready to wrap things up.

Index

Symbols

Packtpub.com

Subscribe to our online digital library for full access to over 7,000 books and videos, as well as industry leading tools to help you plan your personal development and advance your career. For more information, please visit our website.

Why subscribe?

- Spend less time learning and more time coding with practical eBooks and Videos from over 4,000 industry professionals

- Improve your learning with Skill Plans built especially for you

- Get a free eBook or video every month

- Fully searchable for easy access to vital information

- Copy and paste, print, and bookmark content

Did you know that Packt offers eBook versions of every book published, with PDF and ePub files available? You can upgrade to the eBook version at packtpub.com and as a print book customer, you are entitled to a discount on the eBook copy. Get in touch with us at customercare@packtpub.com for more details.

At www.packtpub.com, you can also read a collection of free technical articles, sign up for a range of free newsletters, and receive exclusive discounts and offers on Packt books and eBooks.

Other Books You May Enjoy

If you enjoyed this book, you may be interested in these other books by Packt:

Microsoft Power Apps Cookbook - Second Edition

Eickhel Mendoza

ISBN: 978-1-80323-802-9

- Learn to integrate and test canvas apps
- Design model-driven solutions using various features of Microsoft Dataverse
- Automate business processes such as triggered events, status change notifications, and approval systems with Power Automate
- Implement RPA technologies with Power Automate
- Extend your platform using maps and mixed reality
- Implement AI Builder's intelligent capabilities in your solutions
- Extend your business applications' capabilities using Power Apps Component Framework
- Create website experiences for users beyond the organization with Microsoft Power Pages

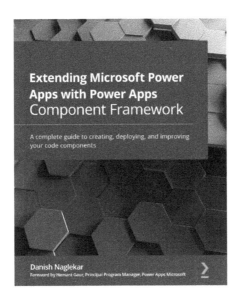

Extending Microsoft Power Apps with Power Apps Component Framework

Danish Naglekar

ISBN: 978-1-80056-491-6

- Understand the fundamentals of Power Apps Component Framework
- Explore the tools that make it easy to build code components
- Build code components for both a field and a dataset
- Debug using test harness and Fiddler
- Implement caching techniques
- Find out how to work with the Dataverse Web API
- Build code components using React and Fluent UI controls
- Discover different deployment strategies

Learn Microsoft Power Apps - Second Edition

Matthew Weston

ISBN: 978-1-80107-064-5

- Understand the Power Apps ecosystem and licensing
- Take your first steps building canvas apps
- Develop apps using intermediate techniques such as the barcode scanner and GPS controls
- Explore new connectors to integrate tools across the Power Platform
- Store data in Dataverse using model-driven apps
- Discover the best practices for building apps cleanly and effectively
- Use AI for app development with AI Builder and Copilot

Packt is searching for authors like you

If you're interested in becoming an author for Packt, please visit authors.packtpub.com and apply today. We have worked with thousands of developers and tech professionals, just like you, to help them share their insight with the global tech community. You can make a general application, apply for a specific hot topic that we are recruiting an author for, or submit your own idea.

Share your thoughts

Now you've finished *Automate Testing for Power Apps*, we'd love to hear your thoughts! Scan the QR code below to go straight to the Amazon review page for this book and share your feedback or leave a review on the site that you purchased it from.

https://packt.link/r/1803236558

Your review is important to us and the tech community and will help us make sure we're delivering excellent quality content.

Download a free PDF copy of this book

Thanks for purchasing this book!

Do you like to read on the go but are unable to carry your print books everywhere?

Is your eBook purchase not compatible with the device of your choice?

Don't worry, now with every Packt book you get a DRM-free PDF version of that book at no cost.

Read anywhere, any place, on any device. Search, copy, and paste code from your favorite technical books directly into your application.

The perks don't stop there, you can get exclusive access to discounts, newsletters, and great free content in your inbox daily

Follow these simple steps to get the benefits:

1. Scan the QR code or visit the link below

https://packt.link/free-ebook/9781803236551

2. Submit your proof of purchase
3. That's it! We'll send your free PDF and other benefits to your email directly